Games and Strategies for Teaching U.S. History

by Marvin Scott

User's Guide
to
Walch Reproducible Books

As part of our general effort to provide educational materials that are as practical and economical as possible, we have designated this publication a "reproducible book." The designation means that the purchase of the book includes purchase of the right to limited reproduction of all pages on which this symbol appears:

Here is the basic Walch policy: We grant to individual purchasers of this book the right to make sufficient copies of reproducible pages for use by all students of a single teacher. This permission is limited to a single teacher and does not apply to entire schools or school systems, so institutions purchasing the book should pass the permission on to a single teacher. Copying of the book or its parts for resale is prohibited.

Any questions regarding this policy or request to purchase further reproduction rights should be addressed to:

Permissions Editor
J. Weston Walch, Publisher
321 Valley Street • P.O. Box 658
Portland, Maine 04104-0658

1 2 3 4 5 6 7 8 9 10
ISBN 0-8251-03772-1

J. Weston Walch, Publisher
P. O. Box 658 • Portland, Maine 04104-0658
Printed in the United States of America

Contents

INTRODUCTION

To the Teacher

An American history class can take many forms. It can be 30 sophomores milling about in a Civil War simulation or a single student entering data into a computer. It can be as physical as hefting a muzzle-loading rifle, or as intellectual as a debate on a constitutional issue. It can include games—some simple enough so that any student can play, even if that student can't read—and others that challenge the creativity of the best. It can be economic, social, or political history. And, laced through it all can be an emphasis on reading and writing.

An American history class can be all of these and more because it has to motivate, inform, and evaluate one of the most frustrating, apathetic, foolish, cheerful, diligent, wise, and charming creatures on earth: the American history student. Other people reading the above will think that I scrambled my adjectives, but you and I know better, because we are American history teachers. We know that students come in all shapes, sizes, and attitudes. During the year those shapes, sizes, and attitudes can change—in seconds or months, depending on each student. It is our job to take these volatile creatures and somehow improve their understanding of our country's development. On top of that, we are expected to make our students better citizens, and improve their skills in reading and writing. We are expected to do all this in roughly 45 minutes a day.

The question is, how can any normal human being do all that? There is no easy answer. All kinds of ideas look great in theory but fall apart in practice. Remember television teaching, programmed instruction, team teaching, and modular scheduling? Where are they now? Mostly on the scrap heap. In certain limited situations they worked, but they were oversold. That is why I'm not promising any miracles—well, not a whole lot of them, anyway. Maybe somewhere in here is an idea that will motivate that impossible student. Notice the "maybe."

This is a collection of teaching strategies that I have used in classes over the past 35 years. They range from large group activities, like the constitutional convention, to assignments and even a whole course for one student. There are quite a few games here because I like to design games, and I find that my students enjoy them. They get involved and excited playing the stock market or managing a railroad. Most of the games are simulations, but there are also some other kinds. "Discovering America" is a combination of quiz show and board game. Simple board games about how a bill becomes a law and how a constitutional amendment is passed are also included. All games have been either field-tested in classes or adapted from games that have.

No matter how interesting games are, they can't be the whole class, so I have included a selection of other teaching strategies. I hope that you will find some helpful ideas about teaching reading and writing or some creative topics for oral reports in those chapters. Don't overlook the best educational exercise of all—debate. I've tried to distill 30 years of debating experience and apply it to American history, so even a total beginner should be able to set up and run classroom debates.

Throughout, I have tried to make the process clear enough for you to manage with a minimum of detail work. The rules are set up to be copied if needed, and several transparency masters are included. All you have to do is read the chapter and make the copies or transparencies, and then the game or activity is ready for your classes.

You will undoubtedly notice some redundancy, that is, two or more activities about the same topic. This is particularly true of political topics. In such cases, you can choose the activity that best fits your classes. That may be a revision, an adaptation, or even a combination of activities. After all, you are the expert on your classes, in the best position to decide how my activity should be managed in your classroom.

National Standards

I have reviewed the most current National Standards for History and believe that much of this material will help reinforce key concepts and eras delineated there. We have finally gotten away from "discovering" America and recognized that the continent was populated before the Europeans saw it. Also, African history is acknowledged, which is long overdue. The outline includes the history of Native Americans, African Americans, and women throughout. The standards include skills in historical thinking, which create a need for activities that stimulate student thinking. I have tried, I hope successfully, to provide many such activities.

About the Author

At this point you are probably wondering whether you should buy a used lesson plan from this person. Does he know what he's doing? Will his ideas work in a real classroom? Here are a few facts that I hope will reassure you.

Marvin Scott teaches in Ames High School, Ames, Iowa, and:

- was a full-time teacher from 1960-1997
- has taught U.S. history for 25 years
- has published four books on teaching
- has coached debate and forensics for 30 years
- has been a judge at national speech and debate contests
- is a member of the Gaming Manufacturers Association's (GAMA) Committee on Education
- has attended and sometimes presided over caucuses in his political precinct since 1966
- has served several terms as precinct committee person, working on election campaigns
- has written articles for *Games and Education*, GAMA's newsletter on education
- has served as editor of *Grub Street*, the newsletter of the Ames Education Association, for fifteen years
- has served on an Ames High School committee on writing across the curriculum
- is serving as a member of the Ames Community Schools' Social Studies Cabinet
- has made presentations in small-group sessions of the Conference of the Iowa Council for the Social Studies on "War Gaming: Simulating Military History" and "Teacher as Writer, Writer as Teacher"

Some Free Material

With shrinking school budgets, material can be a real problem, but teachers can still get some things at the best price: free. One of the simplest ways to do this is to write your representative or senator. Their addresses follow:

The Honorable ____________________
United States Senate
Washington, DC 20510

The Honorable____________________
U.S. House of Representatives
Washington, DC 20515

The tradition of service to constituents has apparently been a victim of change. My representative was considering running for governor, so maybe that had something to do with it. In earlier years I had written to my representative and received some very useful pamphlets. This time, there was no response, so I called his Washington office. My call was returned, and a few days later a fat envelope arrived by Federal Express. I couldn't help wondering why a member of Congress was not using the U.S. Postal Service. The envelope contained quite a bit of paper, but little of it was useful. There was a *Consumer Information Catalog* that listed some free publications and *A Guide to U.S. Government Information*. Maybe you will have better luck with your representative.

You can get free single copies of a pocket-sized edition of the Constitution while they last from National Constitution Center, The Bourse Building, 21 South Fifth St., Suite 560, Philadelphia, PA 19106.

The Federal Reserve Bank in my district provides a wide range of background material on the Federal Reserve. I picked up my copies at a convention. Their address is Federal Reserve System, 20th St. & Constitution Ave. N.W., Washington, DC 20551.

The Game Manufacturers' Association (GAMA) sponsors a semiannual publication called *Games and Education* with articles by teachers on using games designed for the commercial marketplace in class. Several people are using games to teach history. Some manufacturers give a discount to teachers buying games for educational use, and sometimes they even offer a game for free. *Games and Education* is free to teachers. Contact David Millians, Paideia School, 1509 Ponce de Leon Ave., Atlanta, GA 30307. Current and back issues of *Games and Education* are available at http://rpg.net on the World Wide Web.

Inexpensive Material

The U.S. government publishes a huge collection of materials on U.S. history and sells it cheaply. When I wrote for the latest catalog, they sent a 15-page 1997 Subject Bibliography Index. Among the entries are "Civil War," "American Revolution," "Civil Rights and Equal Opportunity," and "Presidents." With the index came bibliographies on "Civil War" and "American Revolution." In the "American Revolution" bibliography, I was surprised to see some great posters still available. For the bicentennial, the U.S. National Park Service and *The Times* of London joined to produce a series of posters on the American Revolution. They are big (29" × 39"), colorful, and packed with information. I own part of the set. If you teach the Revolution, they are excellent sources. Here is a list complete with stock numbers and cost as of this printing:

American Navies 1775–1783
024-005-00626-5 $8.00

British Navy 1775–1783
024-005-00641-9 $9.00

British Redcoat 1775–1783
024-005-00636-2 $8.00

Continental Soldier in the
War for American Independence
024-005-00622-2 $7.00

London 1776
024-005-00632-0 $8.50

Philadelphia 1776
024-005-00625-7 $9.00

Compared to the cost of posters in the 1990's, this is a terrific value. To find out more about these and other bargains, contact: Superintendent of Documents, P.O. Box 371954, Pittsburgh, PA 15250-7954 or Washington, DC 20401; phone (202) 512-1800; fax (202) 512-2250 or (202) 512-1716. You can also access the Internet site http://www.access.gpo.gov/su_docs.

Note on Bibliographies

In preparing the bibliographies at the ends of chapters I have been guided by two principles. Many books are included because they are very useful background for the activities described in the chapter. Others are included because I drew either facts, ideas, or inspiration from them. They may also be worthwhile reading, but they are not essential. I have also included addresses for a few Internet sites that may be worth exploring. Be advised that the Internet changes quickly. I visited these sites in the summer of 1997, but they could be obsolete by the time you read this. My annotations should help you select which bibliographic materials, if any, might be useful.

Bibliography

Nash, Gary and Charlotte Crabtree. *National Standards for History Basic Edition.* Los Angeles, California: National Center for History in the Schools, 1996.

Chapter I

Exploring America

Competition is a powerful motivator. When I tested this game in July, school was out, but my test players all insisted that they have a chance to read the textbook. How often do students insist on reading the text? This game is a race to explore America. Three teams—English, French, and Spanish—try to advance their tokens along a course taken by explorers. They progress by answering questions about explorers and exploration. Playing time is roughly half an hour.

Preparation

You may find that the rules are so simple that you don't need to distribute them. You do need to prepare the map transparency and locate some markers. Coins, tokens, or washers will do nicely. I suggest duplicating questions for students to use as a study guide. They are set up so that you can mask the answers with a strip of paper. Check your textbook to see if it provides answers to these questions. If not, you can send students to encyclopedias.

Orienting the Class

First you need to create teams. If your class sits in rows, you can make the rows different teams. You could also have students just count off one, two, three. Once teams are formed, you should briefly describe the game, stressing that all students need to prepare carefully: If a single student on a team misses a question, it hurts. This should create some peer pressure, and perhaps even lead to some peer tutoring. The next day, you are ready to play the game.

A Sample Game

Selecting the first team to answer questions and assigning countries to teams is purely an arbitrary process. I use age of team leaders, oldest first. Height or picking a number between 1 and 100 are other possibilities. Then, flash the transparency on the screen with each team's marker resting on the home country. As you ask questions, each right answer advances the team's token. My players found the first few questions easy but later questions much harder, so I resorted to a few hints to make things easier. On Question 23, for example, I hummed "The Marines' Hymn" hoping the student would think of "the halls of Montezuma" and get "Montezuma." It didn't help.

Questions 10 and 12 have multiple answers. You can have teams take turns providing one answer each if you like. You can add a few questions, or delete some that are too hard. In my test game, all three teams managed to make it to America, but nobody was able to build colonies. The optional rule about answering missed questions would have raised scores considerably.

Debriefing

Strictly speaking, there is no need for a debriefing, but there are some possible activity assignments that could launch the game. In order to get Europe on the map, I had to move it closer to North America. Please point out this distortion of the map. Because of the need to make wide courses, the routes of the explorers are only crudely indicated. The Spanish route does not do justice to Columbus's four voyages. For the French route, I had to move a couple of islands slightly to make room down the St. Lawrence. The English route is not much better. Perhaps your students could draw a better map?

Name ______________________________

Date ______________________________

EXPLORING AMERICA

THE RULES

1. The teacher will divide students into teams: English, French, Spanish.
2. The teacher will ask questions of a student from each team on a rotating basis.
3. For each correct answer, the teacher will advance the marker of the team along the course that team's explorer followed. The first team to reach America wins.
4. *Optional rules for missed questions:*

   The teacher may permit the next team in turn to answer missed questions. If they are unable to answer, the third team will be given a chance. Answering a missed question does not replace the chance to answer questions in a team's regular turn.
5. *Optional rules for colonizing:*

   Once teams have landed in America, the teacher may permit them to set up colonies. They can take over any square by answering a question correctly, subject to these conditions: They must be either in possession of an adjoining square or on the last space of their sea voyage. The Spanish team may, if they choose, start colonization with the square at the tip of Florida.

Name ______________________________

Date ______________________________

QUESTIONS

Armada	1. The great Spanish fleet defeated by the English was called the ______________________.
Renaissance	2. The time of "new birth" in Europe filled with energy and curiosity is the ______________________.
Crusades	3. From 1096 to 1272 the Europeans battled the Moslems in the ______________________.
Pepper	4. The product from India called "black gold" is ______________________.
Monopoly	5. When a business has complete control of the supply of a product, it has a(n) ______________________.
Prince Henry the Navigator	6. The Portuguese ruler who encouraged exploration was ______________________.
Africa	7. The continent the Portuguese first contacted was ______________________.
Bartholomeu Dias	8. The Portuguese explorer who was the first European to round the southern tip of Africa was ______________________.
Vasco da Gama	9. In 1498 the Portuguese explorer who made the voyage to India was ______________________.
Nina, Pinta, Santa Maria	10. Name one of Columbus's ships. (May be repeated.) ______________________.

Name ______________________________

Date ______________________________

QUESTIONS (continued)

Vinland the Good	11. In the year 1000, the Vikings landed in America and called it ______________________________.
Inca, Aztec, Maya	12. Name an Indian group with a highly organized, complex culture. ______________________________.
Bering Sea	13. By what route did the Indians come to America? ______________________________.
Vasco Nuñez Balboa	14. The first Spaniard to see the Pacific Ocean was ______________________________.
Ferdinand Magellan	15. The leader of the first circling of the earth was ______________________________.
Pedro Alvares Cabral	16. The explorer who first landed in what would become Brazil was ______________________________.
Amerigo Vespucci	17. The explorer whom America is named after was ______________________________.
Ponce De León	18. The explorer who first sailed along Florida's coast was ______________________________.
Hernando Cortés	19. The leader of the conquest of Mexico was ______________________________.
Francisco Pizarro	20. The leader of the conquest of Peru was ______________________________.
Colony	21. A settlement of people in a new land controlled by their mother country is called a ______________________________.

Name ______________________________

Date ______________________________

QUESTIONS (continued)

Conquistadors	22. The Spaniards called people who conquered Mexico, Peru, and so forth, ______________________.
Montezuma	23. The Emperor of the Aztecs was ______________________.
Esteban	24. The black member of a number of Spanish expeditions was ______________________.
Coronado	25. The leader of the expedition that discovered the Grand Canyon, among other things, was ______________________.
De Soto	26. The leader of the Spanish expedition to the Mississippi was ______________________.
John Cabot	27. King Henry VIII of England sponsored his voyages to America. ______________________.
Giovanni Verazano	28. His voyages gave the French their claim to the new lands. ______________________.
Jacques Cartier	29. The French-sponsored explorer who helped establish French claims to Canada was ______________________.
St. Lawrence	30. The name of the gulf and river Cartier explored was ______________________.

Name ______________________________

Date ______________________________

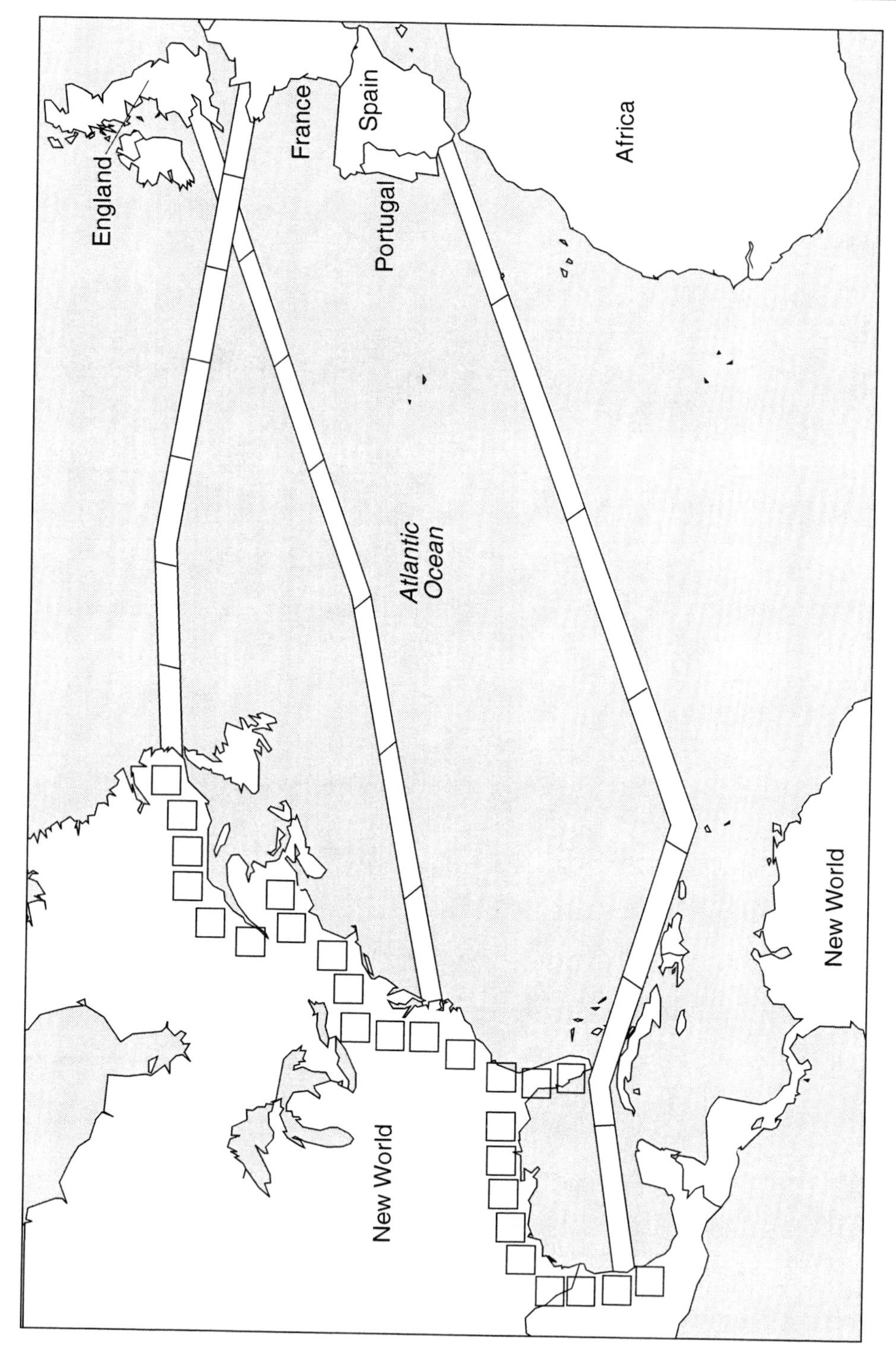
England
France
Spain
Portugal
Africa
Atlantic Ocean
New World
New World

CHAPTER II

A MODEL CONSTITUTIONAL CONVENTION

Study of the Constitution is central to my view of the political history of the United States. The guiding principle of our government is that its power comes from the people through the Constitution. It's a profound idea, the kind that puts teenagers to sleep. Somehow electronic game characters seem more immediate and interesting than a bunch of old duffers meeting nearly two hundred years ago to write about "We the people ...."

The problem for the history teacher is to get those teenagers interested in the Constitutional Convention and in the Constitution itself. The following sample game is an attempt to get students involved in issues touching the Constitution and to give them an experience like the convention. Each student becomes a delegate, and the assembled delegates amend the Constitution. The convention runs two class periods with at least part of another needed for orienting the class.

PREPARATION

You will need a copy of the rules for each student. I suggest that you consider getting another teacher's class to join yours in the convention. Remember, the delegates at Philadelphia were from several states. They may have heard of each other before and met a few times, but many were strangers. Adding a second class to the convention introduces this element. It also adds complications, like coordinating the schedules of the classes and finding a place big enough for the group. Once these details are taken care of, you are ready to orient the class.

ORIENTING THE CLASS

In general, it's probably best if the students have already had some reading and discussion sessions on the Philadelphia convention and the content of the Constitution. Orientation then becomes a matter of handing out the rules, explaining them, and naming the days when the convention will meet. It's probably worth emphasizing, "You are playing yourself. You vote for or against proposals as you feel about them. The time of the convention is now. Try to have reasonable justifications for your votes, and feel free to make speeches to persuade your fellow delegates."

There are a number of ways of preparing amendments. Whether you use an overhead transparency, photocopied handouts, or the chalkboard, what is important is that the system clearly lets delegates know, the nature of the amendment before them.

It's probably a good idea to select a student to act as president. If there is no student with the necessary background, a teacher can play that role. As written, the rules call for a student president backed up by the teacher as parliamentarian. After the orientation, there should be at least a day or two before the convention. Five minute political sessions to draft amendments can be scheduled during regular class time on these days.

A Sample Convention

When the convention started, most problems were solved. The students simply ignored most of the technical details of parliamentary procedure and got down to issues. The clearest memory I have of the convention is that I spent a lot of time hurrying down the hall to the photocopier. The students wrote several amendments, all at the last minute. I did a rush job of copying them while the other teacher supervised the convention.

The amendments covered quite a range. Two stand out in my memory: At first the students were pretty straight-laced about the issues, considering and passing an amendment to elect the president by direct popular vote. But as the convention continued, delegates became less inhibited, passing a final amendment to legalize marijuana. (Shortly after that, we teachers decided that the convention should adjourn.) The convention had been an occasion for energy, enthusiasm, and involvement.

Modifications or Revisions

I have considered and rejected some revisions. Perhaps ideas I have rejected will appeal to you, so I'll explain them and my reasons for rejecting them. Then you can decide if you want to try them.

One of the powerful temptations is to assign each delegate to represent a state, with instructions to vote in that state's interest. If we use "the time is now" framework, I suspect this would be very difficult. Imagine a student in Iowa trying to research the position of Wyoming on direct election of the president. If we moved back to 1787, the task would actually be easier. The original positions of some of the states are on record. Remember the Virginia Plan? My school library has a copy of it, and it is included in *Documents of American History,* in the bibliography below. There are still two problems: Only a few states are documented so well, and how the Constitution's framers resolved issues is too well known. I've always assumed a model convention set up along 1787 lines would recreate the Great Compromise, and so forth, so I haven't tried that model, but maybe you like the idea. Perhaps instead of playing the role of delegates from certain states, students could play certain key delegates. Certainly there are some interesting opportunities there. The student who plays Franklin gets to do the "Rising Sun" speech. "Washington" presides, but what about the students who get the less prominent delegates? If you like the idea, you can probably work out ways to solve the problems. I may not agree with you, but thanks to the Constitution, it's a free country.

Bibliography

Commager, Henry Steele, ed. *Documents of American History.* New York: Appleton–Century–Crofts, 1968.

This is a collection of documents covering the whole sweep of American history. It includes the Virginia Plan, the New Jersey Plan, the Hamilton Plan, and the text of the Constitution.

Farrand, Max. *The Fathers of the Constitution.* New York: Yale University Press, 1921.

This is volume 13 in the *Chronicles of America* series. It is a rather old but standard description of the convention.

Name ______________________________

Date ______________________________

A MODEL CONSTITUTIONAL CONVENTION

THE RULES

1. Each student is a delegate and has one vote.
2. The first order of business is the election of a president for the convention. During this election any delegate may preside.
3. All delegates should have a copy of the Constitution as it is today with them. Your American history text has one in it.
4. If you are planning to submit an amendment, be sure to prepare it as your teacher directs. Write the article number, and then write the article as it would appear if your amendment passed.
5. A teacher will serve as parliamentarian at the meetings.
6. With the exceptions noted above, *Robert's Rules of Order* will be the guide for the meeting.

PLEASE NOTE: As delegates, you must do some planning to run this convention. You can start before the convention by picking your candidate for chairman. Run a campaign for that person. Have some amendments ready when the convention starts. Try to line up support for them before you present them on the floor.

CHAPTER III

THE MINI-STATES OF AMERICA ELECT A PRESIDENT

Wouldn't it be wonderful if every year were a presidential election year? Excitement and drama and wonderful live coverage showing how the system operates to stir students' interest in politics! Teaching politics in a presidential election year is comparatively easy. But the Founding Fathers had more important things in mind than helping out social studies teachers, so three out of four years are not presidential election years.

The Mini-States of America game is my effort to correct this. It brings to the classroom a nomination and election process very similar to that used in the United States. The game can't fully duplicate the excitement of a real campaign, but it gets students involved and produces some useful examples.

The game is designed for classes of 25 to 35 students, but with a little ingenuity, you can scale up or down, depending on the pace you set. It takes 50 to 90 minutes to play. The teacher acts as control and sets the pace. All students are involved in some aspect of the game—as candidates, election officials, delegates, or voters.

PREPARATION

Assuming that your class size is within the range above, all you have to do is prepare one copy of the rules for each student. Read the rules carefully so that you have a plan for how you will handle each stage. Assigning each player a number may require some thought. Will you use a random assignment, or will you select people to be election officials?

Although the player numbers are listed to ensure balance in classes numbering 25 to 35, you will need to adjust the list if your class is smaller or larger. There should be an equal number of heavily Loco-foco and heavily Mugwump states, with one state evenly split between the parties. If you have 15 students, just use 3 states, Neonullus, Alterno, and Gravinia. You can add or subtract Independents to fit the size of your class. If you have more than 35 students (wow!), add Independents to each state until you have accounted for all students. If you want students to be more involved, add members to the majority party in each state or to both parties in balanced states. If you happen to have an even 50, simply double the population of each state. Just remember that you want the presidential election to be in doubt.

You may wish to assign some reading on American presidential elections. The game does not depend on students' being familiar with presidential elections, but familiarity does help them understand the analogy.

ORIENTING THE CLASS

First I distribute the handout and have the students read the introductory paragraph and look through the rules. I walk to the left side of the room and say, "Remember your number," and then point to each student in turn, saying, "one, two . . ." until they each have a number. "Look up your number on the first page. All election officials, hands up." I check to see if I have five. Because somebody always has forgotten his or her number, we have to get that clear. Then I assign states a location in the room. "Neonullus, move to that rear corner; Alterno, the other rear corner. Gravinia, take this spot in the middle; Patina and Irascitur, along the sides there and there." I am reserving the two front corners for the party convention later. The students mill around locating their states and moving desks into five circles. In a minute they are seated. "You have read the rules. Before we start the game, are there any questions?"

"I'm an Independent. Can I go to a convention?"

"No. Only members of the two major parties can be delegates to conventions."

"Could I be a candidate?"

"That's up to the conventions. If they choose to nominate an Independent, that's their business."

"I'm an election official. What should ballots be like?"

"A ballot should have the names of all the candidates on it. Probably it will identify the party of each candidate. Other than that, it can be in any design that state election officials decide to use. The rules show a sample ballot with the presidential candidates, but you do not have to list the presidential candidates. Most states do, but a few just list electors instead. You have that option."

When there are no further questions, we start the game.

A Sample Game

"It's time for state political action. Look at that section in your handout and follow the instructions. You have seven minutes." I circulate among the states to see how they are working. I make a point of not making decisions for students. They have to decide whether their candidate for governor will also be a delegate to the national convention. I will, if asked, point out the possible approaches, but they are the politicians. After six minutes, or when the states appear to be finished with their processes, I move on to national conventions. Notice that I fudged a little on time: Flexibility helps keep things running smoothly.

"You will have roughly 10 minutes for the national convention. The Loco-focos will meet in this front corner. The Mugwumps will meet in the other front corner. If you are not at the convention, check the list to see whether you can be doing something useful. Election officials, now is the time to get the ballots ready." (I act as newscaster in the game. Other teachers may decide to have a student play this role.) The two groups of delegates stand in their respective corners. They usually have some questions.

"Whom can we nominate? Does the candidate have to be a delegate at the convention?"

I answer, "You can nominate anybody you want as long as that person is in your party and willing to run. You want a candidate who can win the election. The candidate probably should not also be running for governor." This leads to a consultation. Since there are only five delegates, they can agree quickly on a candidate. They tell the candidate that he or she has been nominated. I tell the candidate to prepare a short speech to the voters. As soon as both parties have a nominee, I announce, "We will now have a speech on nationwide television by each candidate: first the Mugwump, then the Loco-foco. You each have two minutes."

The candidates' speeches can be quite varied. They range from, "I'm running for president; please vote for me," to a carefully planned speech listing a five-point platform. Plank Five in the platform is "Abolish school lunches," which is not so bad as Plank Four, "Abolish required American history."

My students have always delivered the speeches while standing in front of the class, but it occurs to me that these speeches could really be "televised" if we adjusted the game schedule. The candidates could videotape speeches beforehand and then play them for the class. Or they might even be able to do a live broadcast from the hall or a nearby room that the class views on a monitor. Videotaping would complicate the game a little, but it's an interesting idea. It might get some technically oriented student interested and involved.

The next step in the game is the campaign. I usually allow 10 minutes for milling around as the presidential candidates walk about the room shaking hands and soliciting votes. Length of this stage is a judgment call: If there doesn't seem to be much going on, control can cut it off. At the end, each candidate makes another short speech.

Balloting, the final step, takes little time. Each voter marks his or her ballot and turns it in to the election official, who counts it. I then prepare to tabulate the result. I put a list of the states and three headings on the chalkboard: vote for governor, popular vote, electoral vote. You may want to use the transparency provided. Tabulating the votes sets up the important ideas in debriefing. I first ask the Neonullus election official, "What was the vote for governor?" I put up the response.

"Three Loco-foco and two Mugwump."

"What was your popular vote for president?"

"Three Loco-foco and two Mugwump" is the reply. I note it on the board and move on to each of the other states in turn. After I have tallied results from all five states, I return to the elector of Neonullus. "How do you vote?" In this case the elector is a good Loco-foco and casts all 20 votes for the Loco-foco candidates. This is not the only possible course of action, however. An elector may decide to cast the state's vote for a third candidate not nominated by either party. Electors could also bolt and vote for the other party—unlikely, but it could happen. After I have polled all the states, I total the popular vote and then the electoral vote. Usually the popular vote winner is the electoral vote winner. The winning margin in the electoral vote is usually bigger. But it is possible that the electoral vote winner will not win the popular vote. The resulting table on the chalkboard would look like the one below.

	Governor		**Popular Vote**		**Electoral Vote**	
	Loco-foco	Mugwump	Loco-foco	Mugwump	Loco-foco	Mugwump
Neonullus	3	2	3	2	20	
Alterno	3	2	3	2	15	
Gravinia	2	3	2	3		20
Patina	2	3	3	2	20	
Irascitur	1	4	2	3		25
Totals			13	12	55	45

The Loco-foco candidate for president is the winner. The popular vote is 13 Loco-foco to 12 Mugwump, but the electoral vote is 55 Loco-foco to 45 Mugwump. The Loco-focos won 2 governorships, the Mugwumps 3. Now it's time to discuss the operation of this system and how it compares to real elections in the United States.

Debriefing

There are all kinds of questions that can be raised about this process. In many cases, you can either provide answers or use the questions as bases for short oral reports or papers. You can also select a limited set of questions: I usually focus on the electoral college, for example. How realistic is the game? Time limits and number of players aside, it's pretty realistic. A couple of minor discrepancies could be noted. In a real election, the parties decide how many delegates a state can send to the conventions, and announced candidates would be campaigning for delegates before the convention. (See Chapter XIV, pages 103-105, for tables of delegate votes by states for 1996.) Depending on the class and teacher these preconvention activities could also be part of the game, but they aren't built in. Also, the vice-presidential candidates are not included in the game. The third discrepancy is that in the game the entire electoral vote is cast by one elector; in real life there is an elector for each vote. (See Chapter XIV, page 106, for a table of electoral votes by states for 2000.)

How do conventions select presidential nominees? State caucuses and primaries select delegates. If a candidate is supported by a majority of delegates before the national convention starts, the convention is just a formality. In earlier times, conventions saw dramatic on-site struggles: Lincoln's nomination was aided by a packed gallery; Harding was selected in a "smoke-filled room" in 1920; and the 1960 Kennedy and 1964 Goldwater campaigns were exciting. In 1996, however, both nominations were decided in the state caucuses and primaries, and the conventions merely made them official.

What other things happen at political conventions? Party platforms: What is a party platform? Do presidential candidates accept and adhere to party platforms? How is it decided who will be a party's candidate for vice president? How has this system changed from the original process described in the Constitution? Should it be changed? This may lead to a debate on whether we need a vice president. Discussing conventions in themselves may be worth some time: How did the convention system start? What system did we use before the convention system? Should we replace conventions with a national primary?

Political parties and campaigns are a field for interesting questions. Several students may be wondering, What's a Loco-foco? What's a Mugwump?—excellent historical trivia questions for short oral reports. However, let's get a bit more basic: Where and how did the Democratic Party start? Where and how did the Republican Party start? Have we had other political parties? What happened to the Federalists? The Whigs? What has been the experience of third parties? The Know Nothings? The Free Soilers? The Populists? The Progressives?

The assumption that television would be used in the mini-states campaign could be an opportunity to discuss how television has influenced presidential campaigns. How did candidates campaign before television? How did Washington campaign? Lincoln? Truman? Students might want to explore some famous campaigns in the television era. How did the television debates influence the 1960 campaign? How did Nixon use television in 1968? How have television debates influenced other campaigns? How has television advertising affected campaigning? Should we have rules about use of television by candidates?

In debriefing the game, I focus on the electoral college and its functions and malfunctions. The key questions are these: How does the electoral college work? Can electors vote for somebody other than their party's nominee? How often does this happen? Can the electoral college elect a president with a minority of popular votes? Should we abolish the electoral college? These questions lead to discussing the winner-take-all principle of the electoral college and electoral reform. They focus students on some of the key issues about the presidential selection process, issues that the Founding Fathers discussed and that every American should examine, even if it's not an election year.

BIBLIOGRAPHY

Weinbaum, Marvin G., and Louis H. Gold. *Presidential Election: A Simulation with Readings.* New York: Holt, Rinehart & Winston, 1969. (This is a college-level simulation which started me thinking along the lines that led to the Mini-States game.)

White, Theodore H. *America in Search of Itself: The Making of the President 1956–1980.* New York: Harper & Row, 1982. (A survey of American presidential elections by a journalist who has observed them closely.)

_____. *The Making of a President, 1960.* New York: Atheneum Publishers, 1961.

_____. *The Making of a President, 1964.* New York: Atheneum Publishers, 1965.

(Both of the above provide insights on how a president is nominated and elected. For material on more recent elections, see Chapter XIV.)

Name ______________________________

Date ______________________________

THE MINI-STATES OF AMERICA ELECT A PRESIDENT

THE RULES

INTRODUCTION

Welcome to the Mini-States of America. It is a small country of only five states, but, like the United States, it has two political parties and presidential elections. The two systems are quite similar, except that the Mini-States are much smaller. Even time is smaller in the Mini-States. The entire process of electing a president takes place in less than an hour. In the United States it takes nearly a year. As a member of this class, you are now a citizen of the Mini-States of America. Check your number to see which is your party and state.

Player Number	Party	State
1	Loco-foco	Neonullus
2	Loco-foco	Neonullus
3*	Loco-foco	Neonullus
4	Independent	Neonullus
5	Mugwump	Neonullus
6	Loco-foco	Alterno
7	Loco-foco	Alterno
8*	Independent	Alterno
9	Mugwump	Alterno
10	Mugwump	Alterno
11	Mugwump	Gravinia
12	Mugwump	Gravinia
13*	Mugwump	Gravinia
14	Independent	Gravinia

Name ______________________

Date ______________________

THE MINI-STATES OF AMERICA ELECT A PRESIDENT

15	Loco-foco	Gravinia
16	Mugwump	Patina
17	Independent	Patina
18*	Independent	Patina
19	Loco-foco	Patina
20	Loco-foco	Patina
21	Loco-foco	Irascitur
22	Mugwump	Irascitur
23*	Independent	Irascitur
24	Independent	Irascitur
25	Independent	Irascitur
26	Loco-foco	Neonullus
27	Independent	Neonullus
28	Mugwump	Neonullus
29	Independent	Alterno
30	Mugwump	Gravinia
31	Independent	Gravinia
32	Independent	Patina
33	Loco-foco	Irascitur
34	Mugwump	Irascitur
35	Independent	Alterno

*election official

State	Electoral Vote
Neonullus	20
Alterno	15
Gravinia	20
Patina	20
Irascitur	25

A majority of 51 electoral votes is needed to elect a president

TIME SCHEDULE

The teacher will announce the start and end of each period, lengthening or shortening the periods as needed.

OPENING OF GAME (5 MINUTES)

1. Assign each player a number. The players find their party and state on the list attached.
2. Have players move into groups representing states.

STATE POLITICAL ACTION (7 MINUTES)

Each party must select a candidate for governor, a delegate to the national convention, and a presidential elector. These may all be the same person or three entirely different people, or other combinations may be worked out. How these people are selected is purely a state matter. The party may ask the election official to run a primary in which a secret vote is taken, or they may select people in a state convention.

NATIONAL CONVENTION (10 MINUTES)

1. The two parties hold their national conventions in designated corners of the room. One delegate from each state will attend. Rules of voting and procedure are adopted by the convention. Both parties nominate a presidential candidate. He or she should not be a candidate for governor.
2. A selected player may be asked to act as newscaster, announcing developments at the convention to people not attending.
3. Candidates for governor may campaign in their states if they are not attending the convention.
4. The Loco-foco candidate can appear on "nationwide television" to make a two-minute speech on why people should vote for him or her.
5. The Mugwump candidate can appear on "nationwide television" to make a two-minute speech on why people should vote for him or her.
6. Election officials will prepare official ballots for their states. Design of ballots is their choice.

Name ____________________

Date ____________________

TIME SCHEDULE

THE CAMPAIGN (10 MINUTES)

1. The presidential candidates may move from state to state, trying to win votes.
2. Candidates for governor may campaign within their states.
3. Party members may campaign for their party.
4. Newscasters may broadcast "Gallup Poll" results predicting various election outcomes.
5. Each candidate for president may once again appear on "nationwide television" to make a final appeal for votes.

BALLOTING (5 MINUTES)

1. In each state the voters use ballots provided by the election official.
2. The election official counts the votes and announces the winners for governor and elector. Each state will be give a chance to announce its results to the entire group.
3. The electors in turn announce their votes for president. The candidate with a majority, more than half the votes, wins.
4. The loser congratulates the winner graciously, thanks him or her, and praises him/her as a "great American."

A sample ballot and an election result form are shown on the next page.

SAMPLE BALLOT

MUGWUMP	**LOCO-FOCO**
President	President
☐ Irma Candidate	☐ Als O. Running
Governor	Governor
☐ Hank Polit	☐ Patty W. Heelhorse

Mini-States Election Results

	Governor		Popular Vote		Electoral Vote	
	Loco-foco	Mugwump	Loco-foco	Mugwump	Loco-foco	Mugwump
Neonullus						
Alterno						
Gravinia						
Patina						
Irascitur						
Totals						

CHAPTER IV

THE CONGRESS GAME

The headlines tell the story, "Bill Passes Senate," "President Vetoes Bill," or "Senate Committee Holds up Bill." It's enough to make the average high school student wonder who this "Bill" is.

Getting students interested in "Bill" and how he becomes a law is a challenge we all face when teaching the Constitution. Certainly, students need to know about Congress and how it works. The following board game is designed to familiarize students with the route a bill takes to become a law. It takes one class period, roughly 45 minutes, for the entire process, from explaining the rules through debriefing.

PREPARATION

To prepare this game, you need to make enough copies of the playing sheet for your classes. I usually issue them to groups of 4, so if you have a class of 36, 9 copies will be enough. You will also need an equal number of small cardboard triangles about half an inch long. You can use cereal box cardboard, index cards, or any similar material you have handy.

If you have an overhead projector, you should also make a transparency of the playing sheet. Now, try a few tosses of the triangle. It isn't always easy to drop the triangle so it hits the square of numbers. Practice a few times before you demonstrate this for the class.

If students are going to use tokens to represent their bills in the game, you need to supply these also. Instead of tokens, my students fold up a piece of paper into a small lump to use as a bill. For those of you who must be neat, poker chips, pennies, or washers will do nicely. Once these supplies are together, it's time to start the game.

ORIENTING THE CLASS

Getting the class organized is simple. I say, "Please remember your number," and start counting: one, two, three, four; one, two, three, four. Then I present the "ones" in each group with a playing sheet and a triangle. If you are using tokens, this would be the time to distribute these as well.

Next I do a demonstration. As I turn on the projector, I say, "You will be playing on a sheet that looks like this. Your object is to take a bill and move it through Congress to become a law. To represent your bill, take a scrap of paper and put your name on it. The fold it up into a small square. Since all bills start in the House of Representatives, you start your paper in House Committee." I place a small square of paper on the box marked "House Committee." "To find out what happens to the bill in House Committee you pick a number using the square in the lower right corner of the playing sheet. To do this, drop a triangle on the square. The number on the point of the triangle is the number you use."

At this point I hold up the triangle. Then I position it an inch or two above the square on my transparency master and drop it. My hand

and the triangle, which flutter, unpredictably, are projected as black shadows on the screen. Let's assume the triangle lands as in this diagram. "The number is two. In House Committee, two means the committee votes to kill—bill dies. So this time the bill will not go any further.

1	2	3	4
4	3	2	1
3	1	4	2
2	4	1	3

"For purposes of an example, I'm going to try again." This time the triangle hits in the three box: "Three means bill passes. Bill goes to House Rules Committee." I move my paper square to the box marked House Rules Committee. I drop the triangle, and it lands on a square with a three. "Three means the bill goes to the House. By now you have the idea. Before you start, please notice one thing: If a bill is amended in the Senate, it must go to Conference Committee. Get your groups together and start playing. Select a secretary, and record how many bills die and how many become laws. Any questions? If not, go!"

1	2	3	4
4	3	2	1
3	1	4	2
2	4	1	3

A Sample Game

The students quickly move their desks into clusters of four and begin to fold up "bills" and drop triangles. Some are trying to control the triangle, a tricky process indeed! Sometimes, students have trouble getting the triangle even to stay on the square of numbers, so I suggest holding the triangle closer to the paper. I keep moving among them to check how they are playing. "Are you taking turns moving your bills? Each of you should get a number in turn. Don't let one person just keep playing his or her bill. Take turns. Who is secretary in the group?"

Usually students have a few questions as they play: "My bill passed the House. Where do I go now?"

"Now you move to Senate Committee," I tell them.

A student will often ask, "How do I know if a bill needs to go to Conference Committee?"

I reply, "If it has been amended either in the Senate Committee or on the Senate floor, it should go to the Conference Committee."

Mostly the students are cheerful, dropping triangles and moving markers. They seem to enjoy the process. I let them play for about half an hour. I want a short time, about ten minutes, for debriefing.

Debriefing

"Time to stop now. I want each secretary to tell me how many bills in your group died and how many passed." Each secretary reports, and I note the numbers for each group on the chalkboard. Then I add up the number of bills killed compared to those passed. "What is the pattern you see?" I ask.

"More bills die than are passed," someone volunteers.

"In fact, a larger percentage die in the real Congress than in this game," I reply.

I emphasize several other points in the debriefing: How many members are there in the House? How many members in the Senate? Which house of Congress has a rules committee that screens all bills? Which house of Congress permits a filibuster? What is a filibuster? How likely is Congress to override a presidential veto? Depending on the time available, we can chew on these and other questions. Somewhere along the way, I work in the idea that the Senate and the House each have several committees. House Committee and Senate Committee in the game stand for all the possible committees that a bill could pass through.

Perhaps the next day, I give the following quiz.

Key. 1,S; 2,H; 3,H; 4,B; 5,B; 6,S; 7,B.

Name ______________________________

Date ______________________________

CONGRESS QUIZ

Place the correct letter before the statement:

H = House of Representatives, **S** = Senate, **B** = Both

_________ 1. Has 100 members.

_________ 2. Has a rules committee that screens all bills on the way to the floor.

_________ 3. Has 435 members.

_________ 4. Has the power to amend a bill.

_________ 5. Must pass a bill if it is to become law.

_________ 6. Has a rule that allows extended debate (sometimes called filibuster).

_________ 7. Must vote two thirds or more to override a veto.

Name ______________________

Date ______________________

THE CONGRESS GAME

THE RULES—PLAYING SHEET

How a Bill Becomes a Law

House Committee

1. Chairman refuses to consider. Bill dies.
2. Committee votes to kill. Bill dies.

3, 4. Bill passes. Bill goes to rules committee.

House Rules Committee

1. Refused a rule. Bill dies.
2. Delayed. Stays in committee.

3, 4. Given rule. Bill goes to House.

House Floor—435 Members

1. Bill amended and passed. Bill goes to Senate.
2. Bill killed.

3, 4. Bill passed. Goes to Senate.

From Conference Committee

1, 2, 3, 4. Bill passed. Goes to Senate.

Veto

2,3,4. Fails in 2/3 vote. Bill dies.

1. Passes 2/3 vote. Bill goes to Senate.

Senate Committee

1. Bill amended. Stays in committee.

2, 3. Bill passed. Goes to Senate.

4. Bill voted down. Dies.

Senate Floor—100 Members

1. Filibuster. Bill stays on floor.
2. Bill amended. Bill stays on floor.
3. Bill voted down. Dies.
4. Bill passes.

If amended to conf. committee. 1, 2, 3, 4. Bill passes. Goes to president.

Veto

1, 2, 3. Fails in 2/3 vote. Bill dies.

4. Passes 2/3 vote. Bill becomes law.

Conference Committee

1, 3, 4. Compromise reached. Bill goes to House floor.

2. No compromise.

President

1, 2, 3. Signs bill. Becomes a law.

4. Veto. Return to House.

1	2	3	4
4	3	2	1
3	1	4	2
2	4	1	3

CHAPTER V

CONGRESS: A LARGER SIMULATION

There are some classes that seem to have a glow about them, at least as I look back on them. At the time, they were just another bunch of teenagers—a little more outspoken than most, with lots of enthusiasm. One of these kinds of classes chose to run a model Congress. The simulation was a lot of fun for them and for me. We set up a simplified House of Representatives, and the students wrote bills to be processed through it. Then the House adjourned, and we set up a Senate, and the bills continued on their way.

There was much horse trading to get bills through, a little horsing around, and some solid study of Congress. Each student wrote two short papers about Congress during the unit. We could have done a conventional unit on Congress, walking through the textbook or study guide, but the class chose this method instead. Because this simulation takes several class periods, you may want to offer the kind of trade I gave my class. You can either do it the conventional way, or we can do this. It's very important that the students want to do the simulation.

PREPARATION

Once the class has decided that they want to do the Congress simulation, you start doing clerical work, making a copy of the schedule and rules on pages 28 and 29 for each student. You may want to look at my list of committees to see if you want to change them. If the list omits a committee that would deal with an issue that is currently hot, feel free to add it to the list or replace one of my selections. It's a good idea also to have students do some reading about Congress. A little quiz or paper based on the reading helps the reading get done.

You also need to decide the party strengths in the two houses of Congress. In the 105th Congress, the House of Representatives has 222 Republicans and 204 Democrats with 9 others. In a class of 25, that scales down to 14 Republicans and 11 Democrats. The Senate has 54 Republicans and 45 Democrats with 1 other. The same class of 25 could then have 13 Republicans and 12 Democrats. The object is to give the majority party in the real house of Congress a solid majority in the class model. Once you have decided this, you will need to write up little slips with Democrat and Republican on them and put them in a shoe box, hat, fishbowl, or whatever. Students will use these to draw their party affiliations.

You will also need to decide the size of the committees. I suggest that each member of the class have 2 committee assignments. That makes House committees of about 8 members and Senate committees of 10 members. The majority party gets a majority on committees too, making five Republicans and three Democrats on the House committees, and six Republicans and four Democrats on the Senate committees. (You may adjust these numbers.) Once these details are taken care of, you are ready to start Congress.

Orienting the Class

Anticipation is in the air as students receive rules and instructions and start reading and analyzing. After spending a few minutes going over the schedule and rules, have a student pass the box or hat so class members can draw their party assignments. Once everyone has a party assignment,the game can start.

A Sample Game

The classroom I had at the time of the original simulation was ideal for group meetings, with two little rooms next to it and additional space just down the hall. I sent the two parties out of the main classroom to caucus. In a more conventional setup, they could just meet on opposite sides of the room; I've done that several times, and it works well. I made clear that I wanted a written list of party members, a party leader, and committee assignments in about 10 minutes. It's important to keep things moving and to keep the task defined. At the end of the allotted time, I called the political parties back into the classroom. The House met very briefly. In a quick and uncontroversial action, the Republican nominee for Speaker was elected, and committee members from the two parties' lists were elected to their respective committees.

The second day had a little suspense. Would House members produce bills? My class produced a generous supply. A little box on my desk served as the hopper. I accepted any piece of paper as a bill: Most were handwritten on notebook paper, but one or two were scrawled on scrap paper. I assigned a number to each bill and gave it to the chair of the appropriate committee. The committees then met to consider the bills. The process looks organized on paper, but the actual event was a jumble—students thrusting bills at me, committee chairs announcing, "Armed Services Committee meets in the conference room," and clusters of students huddled in meetings. This lasted for the full class period. I tried to monitor committees to see that we had some bills ready for debate, which was difficult because committees were scattered over the room and meeting in rooms and hallways outside as well. The chairman of the Rules Committee became a very important person: If he didn't act, a bill could never reach the floor, and he gloried in his newfound authority. In an exercise of arbitrary power, his committee gleefully drafted rules for each bill. They limited debate on one bill to five minutes. In another case, they opened a bill to unlimited amendments. But they did produce a usable set of bills for house debate.

The next day the House met for debate. Some committees were in session, but most students were present for debate. The bill with the unlimited amendment rule came up. It was a bill to observe a national week for victims of an obscure disease. First it was amended to read "day," then "hour," then "second." Finally, the House voted to kill the bill. Other bills sailed through.

The next day was spent reorganizing the class to become the Senate. Once again, the shoebox of party assignments was passed around. I made no effort to keep students in the same party in both houses, so each party had to start over from scratch in the Senate. Party caucuses produced some new people in positions of power. When the Senate met briefly to get organized, they only had to vote on president pro tem and committee assignments. The Senate is presided over by the vice president, a role I reserved for myself.

The next day the Senate began operation. There were a few new bills, but mostly the committees concentrated on bills the House had already passed. If the teacher wants to try it, the Senate could also be given a proposed treaty for their "advice and consent"; or they could review a presidential appointment. Be sure to have the proper committees available: A treaty is examined by the Foreign Affairs Committee; judicial appointment is investigated by the Judiciary Committee. Recommendations of these committees are reported to the Senate, much as a bill is.

On the sixth and final day of the simulation, the Senate had a floor debate. Two students just could not resist the temptation to filibuster. They got the floor and held it in spite of heckling and pleading. Bills piled up as they spoke on and on. We finally had to adjourn for the day.

Debriefing

An experience like this raises all kinds of questions: Why is the House Rules Committee in such a key position? Why does Congress have a committee system? Does the Senate rule about unlimited debate make sense? Are there any ways that Congress can be reformed? (And the question my students always ask: Do we have to have the test tomorrow?)

Bibliography

Our American Government. What Is It? How Does It Function? 750 Questions and Answers. Washington: U.S. Government Printing Office, 1981.

(This was my standard reference on House and Senate rules, committees, etc.)

How Our Laws Are Made. Washington: U.S. Government Printing Office, 1998.

(This gives a much more detailed explanation of the process.)

The Electronic Model Congress. http://www.col-ed.org/pro/temc.html

(Recommended in *Social Education* 61 (3) April/May 1997. "Citizenship Education and the World Wide Web," Risinger, Frederick C., pp. 223–224.)

Name ______________________

Date ______________________

SCHEDULE

I. **First Day**
- A. Orientation
  1. Distribution of Schedules
  2. Allotment of Roles for House of Representatives
- B. Party Caucuses
  1. Assignment of Committee Memberships
  2. Selection of Party Leaders
- C. Formal Opening of House
  1. Election of Speaker
  2. Election of Committee Members

II. **Second Day**
- A. Submission of Bills
- B. Work of Committees

III. **Third Day**
- A. House Debates

IV. **Fourth Day**
- A. Allotment of Roles for Senate
- B. Party Caucuses
  1. Assignment of Committee Memberships
  2. Selection of Party Leaders
- C. Formal Opening of Senate
  1. Election of Committee Members

V. **Fifth Day**
- A. Submission of Bills
- B. Committee Work

VI. **Sixth Day**
- A. Senate Debate
- B. Debriefing

Name ______________________________

Date ______________________________

CONGRESS: A LARGER SIMULATION

THE RULES

1. Players will draw their political parties. The relative strengths of the parties will reflect the relative strengths of the parties in the current Congress. There will be a separate drawing for each house of Congress.
2. Party caucuses will assign members to committees and elect party leaders.
3. There will be six committees in the House.
   - (a) Armed Services
   - (b) Education and Labor
   - (c) Energy and Commerce
   - (d) Budget
   - (e) Ways and Means
   - (f) Rules
4. There will be five committees in the Senate.
   - (a) Armed Services
   - (b) Labor and Human Resources
   - (c) Budget
   - (d) Energy and Natural Resources
   - (e) Housing and Urban Affairs
5. Committees have the power to pass a bill, amend it, or kill it.
6. Party strength in the committees shall be proportional to party strength in the house of Congress represented. The chairman of each committee shall be of the majority party.
7. The House Rules Committee determines the rules governing a bill. They may deny it a rule, thus denying it a chance to go to the floor. They may set a time limit on debate. They may decide to allow or deny the opportunity to amend a bill.
8. Filibusters are allowed on the Senate floor, subject to the Cloture Rule. A three-fifths vote can shut off debate.
9. The instructor will play vice president, president, and other roles as needed.
10. Members are encouraged to write proposed bills and submit them.
11. In general, the simulation will be run like Congress.

CHAPTER VI

AMENDING THE CONSTITUTION

If you liked the Congress game, this game is for you. Like the Congress game, it teaches the process, step by step, by having players move a counter on a board. Amending the Constitution is more complex than a bill's becoming a law, and the odds against success are even higher. Players may choose the way they will try to get a constitutional amendment proposed. Because several steps are involved, the process takes all the playing time. In my test game, nobody was able to get an amendment passed in the 35 minutes allotted. The game, uses an entire 45 minute class period from orientation through debriefing.

PREPARATION

You will need to prepare one copy of each of the three game boards for each group of four students in your largest class. You will need little paper or cardboard triangles (my version of dice or spinners) to be dropped on games, and you also want some kind of marker for each student. The marker can be a folded bit of paper, a cardboard square, or a small block of wood. In test play, I used coins. You need to make a transparency of the three boards for demonstrations. A few practice drops on the circle are also a good idea. You may even want to play a practice game, to see how long it takes you to get an amendment proposed and ratified.

ORIENTING THE CLASS

The game boards that you reproduce for your classes contain a lot of information. Before students can play, they need a careful introduction running something like this:

"Today we are going to play a game designed to teach you about how our Constitution is amended. It's a little complicated, so you'll need to listen carefully as I give you these instructions.

"The object of this game is to move your counter through the process. Your counter symbolizes a prospective constitutional amendment. You may start your counter/amendment on either of two boards. There are two ways to propose amendments, and you may select either one. In fact, after trying one way, you can even decide to change to the other. If you choose Board I, *Two-Thirds of State Legislatures Apply for Congress to Call a Constitutional Convention,* start your marker on the box *0 States Apply.* Players take turns. When your turn comes, you drop the pointer on the circle at the top center. Then read the directions in the lower right corner. If you get a 5 (set up the pointer on the transparency so that it points to 5), you stay there. If you get an 11 or 12, you advance. When your marker reaches the box labeled House of Representatives, follow the special directions in that box. In general, follow the arrows.

"You may choose to propose by using Board II, *Amendment Proposed by Two-Thirds Vote of House of Representatives and Senate.* In that case you put your marker in the box marked *House of Representatives,* drop the pointer, and follow directions. I just got a ___, so I will [read the directions in the box].

"When and if you get an amendment proposed, go to Board III, *Ratify.* Because Congress decides how an amendment is ratified, the first thing you do on this board is to find out which of the two methods of ratification you will use. You drop a pointer and read the instructions in the paragraph at the top of the page. Your amendment will be ratified either by state legislatures or by state conventions. The instructions for advancing through state legislatures are on the left side of the board, and those for state conventions are on the right. You drop the pointer on the circle and follow the directions for your method of ratification. Now, form groups of four, and I'll pass out the boards, pointers, and counters. You will have roughly half an hour to get a constitutional amendment passed."

A Sample Game

My test-play group settled down to proposing amendments fairly quickly. After a little trouble and some giggling about getting the pointer to hit the circle, three of the four chose the "Two Thirds of State Legislatures Apply" method of proposing. They found it fairly frustrating because they advanced so slowly. After about 10 minutes of play, I reminded them that they could switch to the other way of proposing an amendment. "What are the odds of your making it?" I asked.

"It's difficult," Tonya volunteered. Jamal, who had chosen to have Congress propose his amendment, was analyzing his chances. After 22 minutes, he got his amendment proposed and advanced to ratification. I reminded him to look at the instructions at the top of the sheet to see how his amendment would be ratified. From time to time, I also had to remind other players to read the full instructions.

By now Paul was frustrated by the length of the process, so he moved from the Two Thirds of States Apply board to the Congress board.

Meanwhile, Jamal was moving smoothly through the ratification process. Having discovered that a large segment of the circle has numbers that permit him to advance, he advanced confidently, dropping the pointer in the "sure thing" sector of the circle. Only the time limit of 35 minutes kept him from getting an amendment through. The other players had failed to get an amendment even proposed.

Debriefing

"Why were you having such a hard time getting amendments proposed?" I asked. "What did your amendments have to go through?"

Tonya looked at the board and said, "Oh boy! They had to go through 34 states, the House of Representatives, the Senate, and a convention."

"Do all those handle the amendment?"

"No, just the convention writes amendments. The states and Congress call the convention."

"The game makes that route very hard because never in the history of this Constitution has an amendment been proposed using that method. All of the amendments have been proposed by two-thirds votes of both houses of Congress." Jamal, who had used that method, looked pleased.

"Once you get an amendment proposed, how does it go, Jamal?"

"I had state legislatures ratify, and it went pretty fast. I suppose it's because if you have enough steam to get it through Congress, the states will ratify," Jamal replied.

"That's the idea. You noticed that you could have had state conventions ratify. That has been done only once. The Prohibition repeal amendment was done that way. So far in our history, we have passed 27 constitutional amendments. Five have been proposed and not ratified."

If there is time, there can be further discussion of the rules and how they reflect the real world or fail to do so. Since we have no experience with what would happen if the states applied for a constitutional convention, that whole board had to be based on my best guess. If students want to, they could rewrite it to reflect how they think Congress or the convention would act.

I'm on much more solid ground with the other method of proposing an amendment. The odds shown are a rough approximation of what could happen. Remember, Congress has only proposed 31 amendments since 1788.

The board for ratification reflects experience pretty well, although in only 1 case in 30 has Congress chosen to have conventions ratify.

BIBLIOGRAPHY

The Constitution of the United States of America As Amended. Washington: U.S. Government Printing Office, 1978.

(This is the basis for the game and furnishes material for debriefing. It contains an annotated Constitution and texts of all amendments ever proposed.)

AMENDING THE CONSTITUTION
BOARD I: PROPOSE

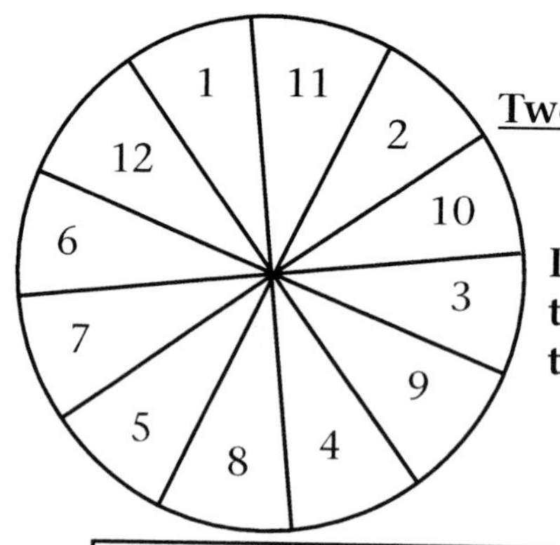

Two-Thirds of State Legislatures Apply for Congress to Call a Constitutional Convention

Drop the pointer on the circle and follow the directions:

34 states apply	29 states apply	24 states apply

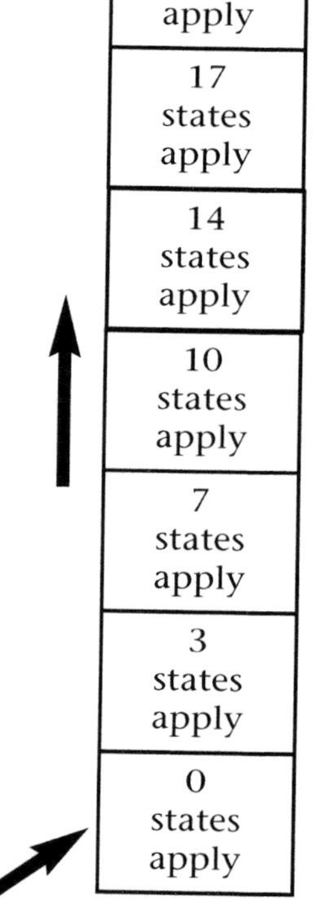

HOUSE OF REPRESENTATIVES
(Vote to Call a Convention)

8, 9, 10, 11, 12—Judiciary Committee votes not to recommend convention call. Start over.
4, 5, 6—Judiciary Committee stalls. Stay in House.
7, 3—House floor vote rejects convention call. Start over.
1, 2—House floor vote calls a convention. Advance to Senate.

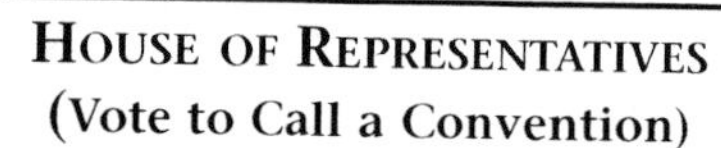

SENATE
(Vote to Call a Convention)

4, 3, 2, 1— Judiciary Committee votes not to recommend convention call. Start over.
6, 5—Judiciary Committee stalls. Stay in Senate.
8, 7—Filibuster against convention call. Stay in Senate.
11, 10, 9—Floor vote rejects convention call. Start over.
12—Floor vote passes convention call. Go to convention.

CONVENTION TO PROPOSE AMENDMENTS

1, 2, 3— Convention decides to write whole new Constitution, but rejects your amendment. Start over.
4— Convention decides to write whole new Constitution including your amendment. Advance to ratification.
5— Convention passes several amendments including yours. Advance to ratification.
6, 7, 8, 9, 10— Convention passes several amendments, but not yours. Start over.
11, 12— Convention decides to make no changes in Constitution. Start over.

START. 1, 2, 3, 4, 5, 6, 7, 8, 9, 10—Legislatures are not interested. Stay on current square.

11, 12—Lobbying is effective. Legislatures vote to apply to Congress. Advance one square.

Name ____________________

Date ____________________

AMENDING THE CONSTITUTION
BOARD II: PROPOSE

Amendment Proposed by Two-thirds Vote of House of Representatives and Senate

Place marker in House. Drop the pointer on the circle and follow the directions:

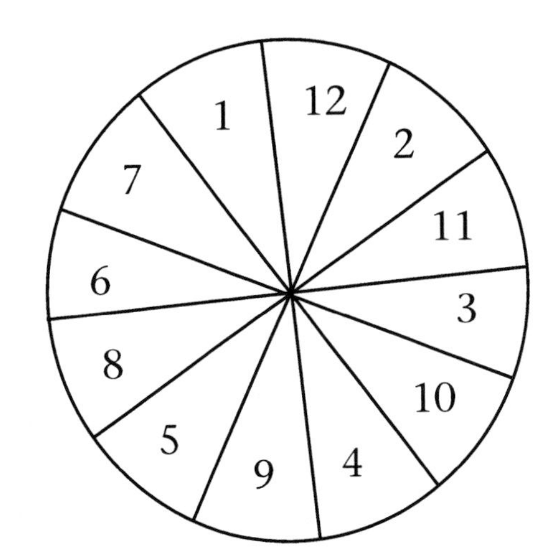

HOUSE OF REPRESENTATIVES
(Two-thirds Vote to Propose Amendment)

1, 2, 3, 4, 5, 6—Judiciary Committee votes to kill the amendment. Start over.
7, 8, 9—Judiciary Committee holds amendment. Stay.
10, 11—House rejects amendment in floor vote. Start over.
12—House gives two-thirds vote. Advance to Senate.

SENATE
(Two-thirds Vote to Propose Amendment)

7, 8, 9, 10, 11, 12—Judiciary Committee votes to kill the amendment. Start over.
5, 6—Judiciary Committee holds amendment. Stay.
2—Senators filibuster against amendment. Stay in Senate.
1—Senate passes by two-thirds vote. Advance to ratification.
3, 4—Senate floor vote fails to pass amendment. Start over.

Name ____________________

Date ____________________

AMENDING THE CONSTITUTION

BOARD III: RATIFY

An amendment must be proposed before being ratified. Congress decides how an amendment will be ratified. Drop the pointer on the circle and follow directions. If you first drop hits 1, 2, 3, 4, 5, 6, 7, 8, 9, 10, 11 start at "0 State Legislatures Ratify." If your first drop hits 12, start at "0 State Conventions Ratify."

THREE-QUARTERS OF STATE LEGISLATURES RATIFY.

4—Amendment is contrary to traditional attitude in the state. Rejected by legislatures. Stay on current square.

5, 6, 7, 8, 9, 10, 11, 12—There is effective lobbying for the amendment. It is ratified by legislatures. Advance one square.

3—Amendment is blocked by a power struggle in the legislature. Stay on current square.

1, 2—There is effective lobbying against the amendment. It is rejected by legislature. Stay on current square.

State Legislatures	State Conventions
0 State Legislatures Ratify	0 State Conventions Ratify
4 State Legislatures Ratify	4 State Conventions Ratify
8 State Legislatures Ratify	8 State Conventions Ratify
12 State Legislatures Ratify	12 State Conventions Ratify
16 State Legislatures Ratify	16 State Conventions Ratify
20 State Legislatures Ratify	20 State Conventions Ratify
24 State Legislatures Ratify	24 State Conventions Ratify
28 State Legislatures Ratify	28 State Conventions Ratify
32 State Legislatures Ratify	32 State Conventions Ratify
38 State Legislatures Ratify	38 State Conventions Ratify

AMENDMENT IS RATIFIED.
IT BECOMES PART OF THE CONSTITUTION.

THREE-QUARTERS OF STATE CONVENTIONS RATIFY.

1, 2, 3—Slates of delegates hostile to the amendment are elected to the convention. Convention rejects amendment. Stay on current square.

4, 5, 6—Slates of delegates friendly to the amendment are elected to conventions. Conventions ratify the amendment. Advance one square.

7, 8, 9—Conventions reject amendment. Stay on current square.

10, 11, 12—Conventions ratify the amendment. Advance one square.

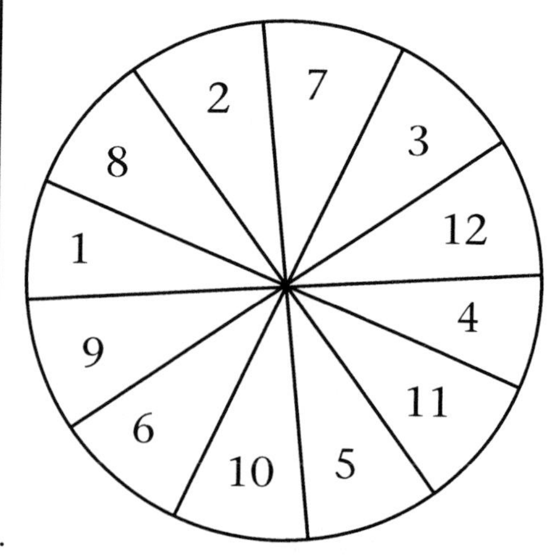

CHAPTER VII

THE CIVIL WAR

The Civil War was America's biggest war until World War II. There were days in the Civil War when more Americans died in battle than in the Revolutionary War, War of 1812, and Mexican War combined. Not only was this a big military event; it was a war about issues central to the existence of the United States: the nature of federalism and relations between the races. The political, social, and economic impact of the Civil War was tremendous, and any Civil War simulation has to be big and complex.

I never had the nerve to try a game about the Civil War in my own classes, but when I asked two colleagues, Kirstin Sullivan and Carolyn Bolinger, what game they would like me to design for them, they both requested a Civil War game. I quickly replied that I really couldn't. But then I woke up one night with an idea. What if I used states as the basic movement. What if a move represented about three months? Let armies fight battles to control states. As a side lost states, it would grow weaker—still too complex, so I went back to sleep. But ideas kept coming. Each side could have a political leader, a strategic commander, and a group of field commanders—that would put students right inside the problems Lincoln faced selecting generals. There could be politics and diplomacy, with ambassadors as well. Students could play the military campaigns and the diplomatic and political decisions. What if Britain supported the Confederacy? What if slaves were freed—flat-out freed—not like the limited Emancipation Proclamation?

In a few days, the ideas went down on paper. My colleagues liked the rough draft of the rules. I reworked them a bit and located a few supplies: We needed maps for the military game, so I located some big blank maps of the United States and modified them. We hunted up dice and stickpins and made copies of the rules for each student. Kirstin and Carolyn tried the game in their U.S. history classes, and students responded quite favorably. The game lasts for about a week and takes a full class period each day. It involves all students, although some have more to do than others. It is a lot of work to set up, but I have tried to provide all the materials you need.

PREPARATION

After some important early decisions about rules and mechanics, there is some routine copy preparation. Selecting a team of umpires may be the most important decision in setting up the game. You need reliable students to keep the game running smoothly. Umpires help you gather materials, like dice, stickpins, transparencies, and markers. They also keep track of the moves, see that rules are followed, and generally keep the game running smoothly. As head umpire, you have the authority to rule on the result of political moves. Your best guess is the answer for questions such as how France would respond to a request for an alliance with the South. (You could also delegate the role of the French government to an umpire.) One umpire should probably be put in charge of the history of the game, keeping records of all political and diplomatic moves, as well as of battles. The history will be valuable during debriefing. You will probably need at least five umpires.

After selecting umpires, divide the rest of the class into two teams, Union and Confederacy. Either designate a political leader for each side or let each group of students select their own leader. You need to prepare materials for the game, including a copy of the rules for each student. You also need to decide how to do the map.

You need one map for each class playing the game. For our trial run, I bought five large blank maps and modified them, adding seaports and combining some states. Carolyn and Kirstin mounted the maps on their bulletin boards and used plastic stickpins to represent armies. Players could tell the military situation at a glance. (Umpires could be assigned to prepare maps and pins.)

For this book I have included a similarly modified outline map of the continental United States. It is like the map we used in the first runs of the game, but it must be enlarged for bulletin board use. Most copiers can enlarge images; or your umpires could project the maps on a paper and trace them. You could also use an overhead projector, which would be a bit simpler. Just make a transparency of the outline map and play out the military moves on the screen, using a series of overlays to indicate the armies. Whatever map system you choose needs to stay for the whole week. Putting the map on a table and using little plastic soldiers to represent armies isn't practical, but it would look good. Soldiers are sold in model shops if you decide to try this, but in my classroom, something would get moved around by the next class, the night school, or the janitor.

The armies need to be fastened in place against accidental shocks yet still be easy for umpires and generals to move. Paper markers will work, too. I have provided some on pages 51–52. You can fasten them to the map with a circle of sticky-side-out masking tape or with pins.

For most purposes, a full map of the United States is unnecessary, since the critical action was confined to the Confederacy and a few nearby states. I have also provided a map of that area. It allows more room than the more complete map to work in the places where most of the action will be.

I recommend that you or the umpires prepare a calendar for display. All it needs to show is the year and season. It should be displayed on a screen or a bulletin board so all the players know what stage they are in. I have provided a master for a calendar at the end of the chapter. You may also want to gather several sets of regular six-sided device. I have also included a random number device. We used dice in our first run. When all supplies are ready, it's time to pass out the rules and orient the class.

Orienting the Class

First, give students some time to read the rules. In the discussion that follows, you need to stress these ideas. The game is intended to be like the Civil War. It will run from the start of fighting to the election of 1864. In general, the more you know about the Civil War, the better you should be able to play the game. Each team has one political leader, one military strategist, and several field commanders. (You may decide to use a committee in the political and the strategic leaders' roles, but for purposes of this example, let's assume you use a single person.) Each role has different responsibilities and takes different abilities.

The political leader will be responsible for political moves (see page 42). Political moves can win or lose the war, and their results are not always clear, to reflect uncertainty about what would have happened. For game purposes, the umpires will decide. If you are asking someone to join you, the French for example, giving them good reasons to do so will raise your chances of success. Notice that you select key people. Choose carefully. You may replace people during the game. Both Lincoln and Davis did.

The military strategist plans and executes the strategy: The Union strategist has a navy to deploy. The Confederate strategist does not, but has to decide if a navy is useful in defending the ports or if armies will suffice. Armies raise lots of questions. Will you attack or defend? Where will you position the most forces? Who will command? Can you measure up to General Grant?

The field commander follows orders and carries out the plan. A good field commander can become a hero, like Stonewall Jackson. A bad one can make a good plan fail.

Let's look at a sample battle. In the summer of 1861, the Union decides to advance out of Maryland and destroy a Confederate force guarding the northern border of Virginia. The Union commander moves one corps into Virginia and declares his or her intent to attack the Confederate army. The Confederate commander elects to defend with a force that also equals one corps. The Confederate marker (stickpins, paper marker, or notation on the transparency) stays in place. The Union marker is moved into Virginia next to the Confederates. At this point, the umpire gives each field commander a single die. Since the Union is attacking, the Union commander must beat the Confederate commander's roll by three. Assume the Confederate commander rolls a 3. The Union commander rolls a 5, not good enough. The Union attacker needed to roll three points more than the Confederate defender. The Union commander decides to retreat to Maryland. The umpire notes that the Union army is now only half as strong because of losses during the attack and the psychological impact of the defeat. In future actions in this campaign, the Union's die rolls will be divided by two, assuming that no reinforcements arrive. If the Union army receives reinforcements in the winter of 1861–62, it will return to full strength. If the battle just sketched seems familiar, that's because it is the battle known as First Bull Run (First Manassas).

In the sample battle, neither side gained control of an enemy state. Political leaders and military strategists need to keep one idea constantly in mind: Control of states means power. If you lose a state, your forces will be weaker. If you gain a state, you gain strength. States are the basis of power. Their farms and factories provide the supplies needed for the war. Their population provides soldiers. Obviously, the two sides are not equal, just as they were not equal in reality.

Playing the Game

After the class has gone over the rules, it's time to let the two sides get together, get organized, and do some planning. If you did not assign roles, you can require each side to give you a roster of who is in each role. The umpire team will also need a copy, and you might want an extra for your files. These meetings have the potential to get a bit noisy: How loud a session you tolerate is up to you.

All of this leads up to step five, when each side places its forces on the maps. The umpires should supervise as the Union field commanders go to the map and place their armies in selected Union states and the Confederate field commanders follow suit in their states. Political leaders and strategists are not allowed to accompany them or communicate with them. Umpires should verify that only the allowed number of corps are deployed. After this is done, the game moves on to summer 1861.

In summer 1861, the two sides can move forces and fight battles. The umpires can decide which side moves first. They can use die rolls, coin toss, or whim to decide order of movement. Umpires resolve battles. Let's say a big Union army, three corps, moves into Arkansas and attacks a defending Confederate army of two corps. The umpires provide two dice to the Confederate commander and three dice to the Union commander. The Confederate commander rolls a 4 and a 3. The Union commander rolls a 2, a 5, and a 6. The attacking Union forces win: They have rolled 13, a number at least equal to the 7 that Confederates rolled plus 3 more for each Confederate corps deployed. The Confederate Army loses a corps. The commander can elect to stay in Arkansas or retreat.

One caution: Early moves often take longer than expected. People are still learning rules and are reluctant to take risks. Keep an eye on the time to avoid ending class in the middle of a move. When class ends, be sure that map and markers are secure so they will remain undisturbed until tomorrow, and check to see that umpires have recorded developments so far. As head umpire, you may need to interpret rules.

If our test games are typical, and I suspect they are, the game will focus on the Union hammering its way into the Confederacy. Political strategy will not see much use. I did step into one class to be greeted by a Confederate team saying, "We're going to win. We got Britain on our side." I don't think they won, however. Political moves call for decisions by the chief umpire. If a side asks the British for an alliance, this could add significant forces to their side. As umpire, I would ask them to make the case for the alliance, listing reasons why the British might support that side, and I would base my decision on how persuasive and historically accurate their arguments were. Also, because communication was much slower in the Civil War era, I would not answer immediately. Political moves are made in the winter of each year, so I would announce Britain's reply after the spring move. This allows some time to think and consult the team of umpires. I might even assign an umpire the task of checking the reasons offered and making a recommendation.

Some political moves have results that could extend over years of play. How would border states react to freeing the slaves? How long would it take to build a railroad to California? Such results could be announced in stages, following the move. One way would be to ask your umpires to publish a newspaper, summarizing events at the end of each season or each year.

During the game students move around a lot and cluster in groups with lots of discussion. Some are more involved than others. Most are focused on the map and how the game is going. When it's time to make moves, there is a traffic jam in front of the map. The trial games used about four days for playing time, another day for setup, and an additional class for debriefing—a total of about six classes.

The Small-Group Game

If the full-class version of the game is not your style, you can set up a series of games to be played in small groups. One player could be the Union, another the Confederacy, and a third the umpire. In this case each of the three students in each group needs a full set of game materials, and the umpire also needs the "Political Moves—Results" table that I have provided. Storing all those game maps overnight would be a challenge. Players could compare how their games went. Aside from talking about friction on their teams, the debriefing would be much the same.

Debriefing

Debriefing is a chance to connect the game to historical reality. The first question I always ask after a game is "In what ways was this game unrealistic?" In the test games, students offered some very perceptive responses: "There was no way to recreate the Mississippi River campaign using gunboats on the river." This is certainly true. The rules just don't cover riverboats. "There was no way to destroy an army." These students clearly wanted to see an army wiped out if it lost a battle. It had not happened that way in most battles, however. I revised the rules so armies could lose strength. Now they can forfeit half of their corps if they lose a throw of the dice. But it's important to notice that forces dwindle as the resource base shrinks. This really was the way to destroy an army—the logic of Sherman's march through Georgia, which destroyed a lot of resources that could have supported the Confederate cause. Also, the Mississippi campaign cut resources from Texas, Arkansas, and Louisiana off from the rest of the Confederacy.

"The Confederacy never had a chance to win. The Union had more of everything." Kristin Sullivan has the perfect answer for this one: "Hey, that's history!" Compare the strengths of the two sides in the real Civil War.

"We didn't know the results of the political moves. The rules told us what could happen, not what would happen." Again, that reflects reality. You find out how it will work out only after you do it.

After dealing with objections, you can move on to some of the following questions: Why didn't you try to negotiate a settlement? (None of our test teams tried.) Why didn't the real governments try to negotiate? Did your side have problems working together? Did your "real" historical counterpart have similar problems? Clearly, the answer to the second question is yes—both sides replaced generals who did not work out. How did the blockade affect the war? Why didn't the Confederacy have a stronger navy? How important were the two capital cities in your game? In the actual Civil War? Notice that Washington and Richmond were rated as equal to two states with respect to maintaining forces. In the real war, both sides made serious efforts to capture the other's capital. How did the slavery issue affect your game? How does this compare with the way it actually happened in history? How did the Native Americans fit into your game? How does this compare with the real situation?

The questions above need to be used selectively. Each run of a simulation produces a different set of events. You will undoubtedly have some ideas you will want to stress. You may want to use the game as a springboard for research projects. You could assign each student the task of researching his or her historic counterpart and comparing what happened in the game with what happened in history. This would be easier to do for political leaders and harder to do for field commanders. The list of debriefing questions could be turned into research projects that could be presented as panel discussions, reports, papers, videos, or computer disks. The Civil War is a huge event, a lifelong study for many people. This game provides a lively classroom experience that can start your students on their own explorations of the Civil War.

Bibliography

Battles and Leaders of the Civil War. New York: Appleton–Century–Crofts, 1956.

(In 1883 *The Century* magazine began a series of articles by participants from both sides of the Civil War. The series was collected into four volumes and published in 1887. This is a reprint: fascinating firsthand accounts, illustrated with original old-fashioned engravings.)

The Civil War. Alexandria, VA: Time-Life Books, 1987.

(This series includes 27 slim volumes, illustrated with maps, pictures of artifacts, and contemporary photos. They are colorful and lively accounts of major aspects of the war.)

Boritt, Gabor S., ed. *Lincoln's Generals.* New York: Oxford University Press, 1994.

Davis, William C. *Duel between the First Ironclads.* Mechanicsburg, PA: Stackpole Books, 1975.

Dupuy, Ernest R., and Trevor N. Dupuy. *The Compact History of the Civil War.* New York: Hawthorne Books, Inc., 1960.

Higginson, Thomas Wentworth. *Army Life in a Black Regiment.* Alexandria, VA: Time-Life Books, 1982.

Internet Civil War homepage: http://members.tripod.com/%7ecsa President/index.html

(This is a role-playing game where players become politicians or members of the armed forces of the Union or the Confederacy.)

Name ______________________________

Date ______________________________

CIVIL WAR GAME

RULES

For the next few days, you will be playing a game designed to simulate the Civil War. The class will be divided into teams, Union and Confederacy. Some of you will be asked to act as umpires. The Union and Confederacy teams will try to use politics, diplomacy, and military force to gain their goal. The Union side is trying to force the Confederates back into the union. The Confederacy is trying to deepen its freedom, protect its territory, and above all, protect the rights of the Confederate states. If the Confederacy can defend their territory until the presidential election of 1864, perhaps a new peace president will be elected.

You will find it is helpful to read a bit about the Civil War as you play the game.

SEQUENCE OF PLAY

1. Pass out rules.
2. Assign students to teams: Union, Confederacy, Umpires.
3. Read and discuss rules.
4. Each team meets to get organized. Note that each team needs people at three levels: political, strategic, and field.
5. Start of the war: Each side announces the location and makeup of their forces and provides a list of players and roles. The Union is first.
6. Summer 1861: Forces move, and battles are fought.
7. Fall 1861: Forces move, and battles are fought.
8. Winter 1861–1862: Political moves. Umpires revise strengths. Strategic orders.
9. Spring 1862
10. Summer 1862
11. Fall 1862
12. Winter 1862–63

SEQUENCE OF PLAY (continued)

13. Spring 1863
14. Summer 1863
15. Fall 1863
16. Winter 1863–64
17. Spring 1864
18. Summer 1864
19. Fall 1864: Game ends.

Role of Umpires

1. Prepare maps, dice, and equipment for play.
2. Observe moves, etc., to see that rules are followed.
3. Keep a record of all actions.
4. Help the game run smoothly.
5. Rule on strength gained or lost by each side.
6. Head umpire (teacher) rules on results of political moves.

THREE LEVELS OF COMMAND

I. Political Leader: the Boss

1. Makes political and diplomatic statements.
2. Sets goals for military.
3. Selects the military strategists and advises them if they wish.

II. Military Strategist

1. Decides the makeup of forces.
2. Assigns field commanders.
3. Issues orders to field commanders.

III. Field Commanders

1. Move army (or fleet).
2. Fight battles.

POLITICAL MOVES

Political Leader

1. Free the slaves. Could add a corps. Could lose one or more states.
2. Use black soldiers. Could add a corps. Could lose one or more corps.
3. Ask Britain for an alliance. Could add 10 fleets and 2 corps.
4. Ask France for alliance. Could add 3 fleets and 2 corps, probably from Mexico.
5. Build a railroad west to California. Could add a state, and therefore a corps.
6. Ask Native Americans to join forces. May add a corps.
7. Replace strategists.
8. Spell out political needs to strategist.
9. Negotiate a settlement with the other side's political leaders.

Players may suggest other moves but will need to support them by presenting research to the umpires. Note that some actions already make more sense for one side than the other. Investigate to see if your side has a chance of success. The head umpire's ruling is final.

MILITARY MOVES

Some strategic moves can only take place in winter. Others may be made at the start of spring, summer, or fall.

Military Strategist

1. Decide makeup of armies and fleets and their locations. (Winter)
2. Restore armies or fleets to full strength. (Winter)
3. Convert corps to fleets or reverse. (Winter)
4. Assign commanders to armies and fleets. (Winter)
5. Issue written orders to field commanders.
6. Consult with and advise political leaders.

Field Commander

1. Move the force (army or fleet).

MILITARY MOVES (continued)

2. Fight battles.
3. Decide to attack or defend.
4. May have orders about all of the above, but may choose to do otherwise at the risk of being replaced.

RULES FOR MOVES AND BATTLES

Moves

Forces move in spring, summer, and fall.

1. All forces start in friendly territory.
2. Land forces move one state, port, or capital (in some cases two merged states) per move. The state they move into must have a common border with the state they are in, with no diagonal moves where corners only touch.
3. An army moving by sea must have a fleet with it. It can land only at a port.
4. A fleet can move to any port in one move unless an enemy fleet is blocking it.
5. No army may advance into enemy territory if there is an enemy army in the state where they are. They may, however, move into a friendly state.
6. An army may move into a state, even if an enemy army is already there.
7. Only one side at a time can occupy capitals and ports.
8. All armies must maintain a supply line to friendly territory. Deduct one from each die rolled as long as an army remains cut off.
9. The Union uses New York or Boston as starting ports for ocean travel.

Battles

1. Battles are decided by throwing dice. Each side will roll as many dice as it has corps (or in naval battles, fleets).
2. Armies may decide to attack or defend. They must tell the umpire what they intend to do at the start of each season.

Name ______________________________

Date ______________________________

RULES FOR MOVES AND BATTLES (continued)

3. If two armies are in the same state and both decide to defend, there is no battle. Neither army can advance into enemy territory. Either or both may retreat to friendly territory.
4. If two armies are in the same state and one decides to attack and the other to defend, the attacker must roll a total at least equal to the defender's plus three for each corps in the defender's force.
   a. If the attacker wins, the defender must retreat, and in future battles in that year, it loses half its strength. If it loses two battles in a year, it is wiped out.
   b. If the defender wins, the attacker may remain in the state, but loses half its strength. If it loses two battles in a year, it is wiped out.
5. If both armies decide to attack, the army that rolls the higher total wins. The loser loses half its strength. If it loses two battles in a year, it is wiped out.
6. When fleets fight, the side that rolls the higher total wins, and the loser is destroyed.

GAINING AND LOSING STRENGTH

1. Both sides have the potential to gain or lose through political moves.
2. Both sides gain or lose forces due to change in control of states, capitals, and ports. Each side starts with the following forces:

Confederacy	
9 states and combined states	9 corps
5 ports	1 corps
Richmond	2 corps
Total	12 corps

Union	
18 states and combined states	17 corps
2 ports	5 fleets
Washington	2 corps
Total	19 corps, 5 fleets

3. Losing a state or combined states will reduce forces by one corps for each state or combination of states lost.
4. Gaining a state or combined states will increase forces by one corps for each state or combination of states gained.

Name ______________________

Date ______________________

Gaining and Losing Strength (continued)

5. If a side loses its capital, it loses two corps.
6. If a side takes the other side's capital, it gains two corps.
7. Umpires will announce changes in control of states, capitals, and ports, along with changes in forces for both sides during the "winter" move, after political moves, and before strategic moves.
8. If a state, capital, or port is occupied by enemy forces, the enemy is credited with the forces that the state, capital, or port can support. These forces are removed from the other side.
9. If there are armies from both sides in a state, the state remains part of its original side until the opposing side wins a battle in that state. Then it moves to the other side.
10. A state generally stays on its original side, but it changes sides if the last army through it was from the opposition.
11. In disputes about who controls a state, capital, or port, the umpires' ruling is final.
12. If a state is cut off from its side, it cannot be counted as part of that side, except that it may support an army inside the state, subject to ruling by the umpires.
13. If the South loses three ports, it loses the five fleets supported by ports. If there are no fleets, it loses one corps instead of five fleets?

Forces: Union, Confederacy

Confederacy = 9 corps

Texas, Louisiana/Arkansas, Mississippi, Alabama, Georgia/Florida, South Carolina, North Carolina, Tennessee, Virginia

Union = 17 corps and 5 fleets

Delaware/Maryland, Rhode Island/ Connecticut, Massachusetts, Maine/Vermont/ New Hampshire, California/Oregon, Kansas, Minnesota/Iowa, Missouri, Wisconsin, Illinois, Kentucky, Indiana, Michigan, Ohio, Pennsylvania, New York, New Jersey, West Virginia

Each unit above supports one "corps" = 50,000 men and is one move.

FORCES: UNION, CONFEDERACY (continued)

South—5 ports. Norfolk, New Orleans, Mobile, Savannah, and Charleston. Each supports one fleet, or all five supports one corps.

Each capital city—Richmond and Washington—supports two corps.

POLITICAL MOVES

The following list of suggested odds for results of political moves is based on my best guess. Teachers may modify them or ignore them. They may be used by the umpires of the small-group games.

POLITICAL MOVES: RESULTS

1. Free the slaves. If Union does it, roll one die. If it comes up 6, Union loses the border state with the fewest Union troops in it. If Confederacy does it, they lose South Carolina.
2. Use black soldiers. If Union does it, roll one die. If it comes up 1, Union loses a corps. Any other number, it gains a corps. If Confederates do it, roll one die. If it comes up 1, 2, or 3, they lose a corps.
3. Ask Britain for an alliance. If Union does it, no change. If Confederates do it, roll a single die. If it comes up a 5 or 6 they get the alliance, unless the Union has abolished, slavery. In that case, the South must roll a 6 to form an alliance with Britain.
4. Ask France for an alliance. Same as Britain above.
5. Build a railroad west to California. Either side can try, but only the Union can succeed. Roll a die. Whatever number comes up is the number of moves it will take to complete and add the California corps.
6. Ask Indians to join your side. Roll one die. If you get a 2 or 4, add one corps.

1861:		Fall	1863:	Spring	Fall
	Summer	Winter		Summer	Winter
1862:	Spring	Fall	1864:	Spring	Fall
	Summer	Winter		Summer	

Name ______________________________

Date ______________________________

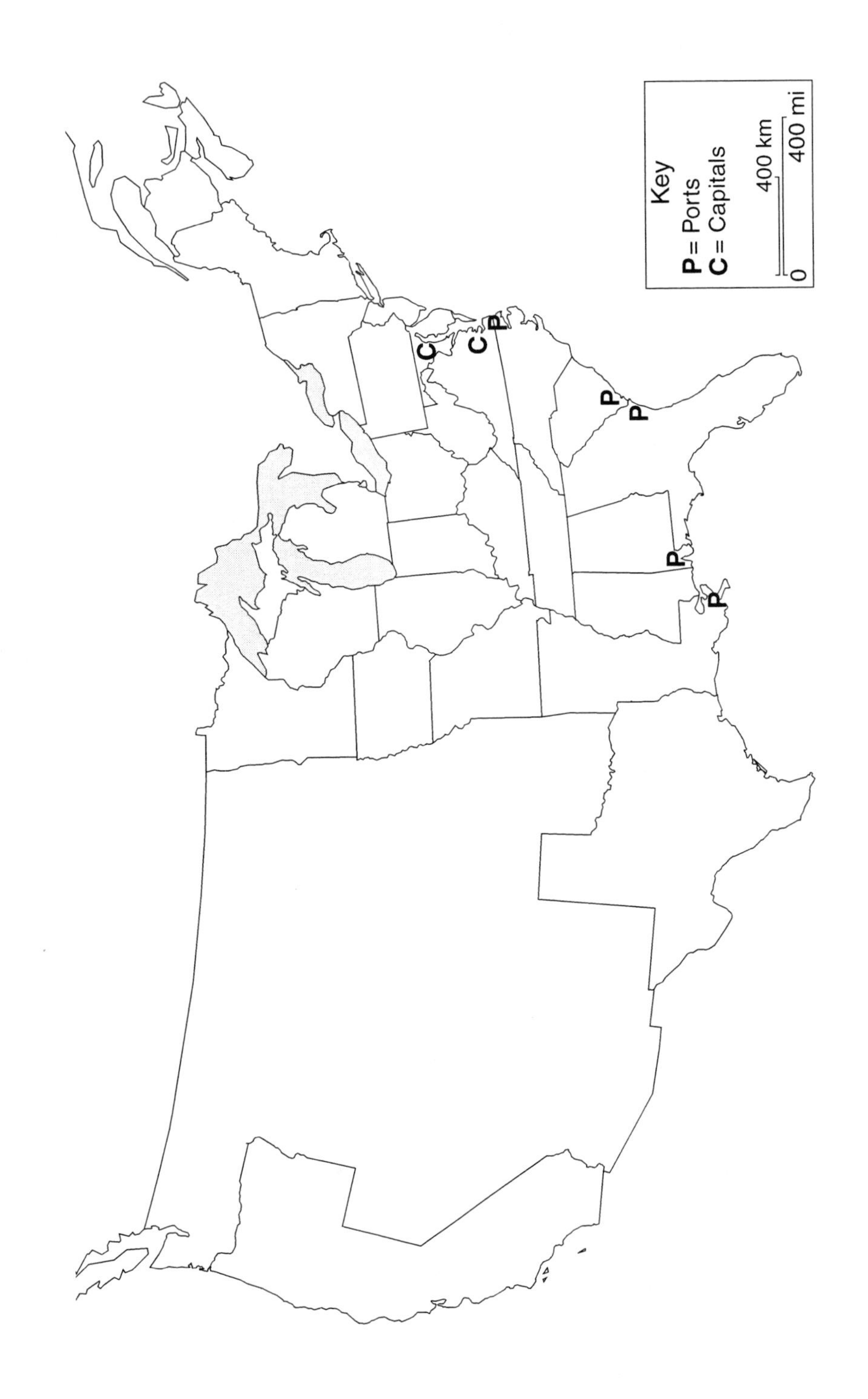

Name ________________________

Date ________________________

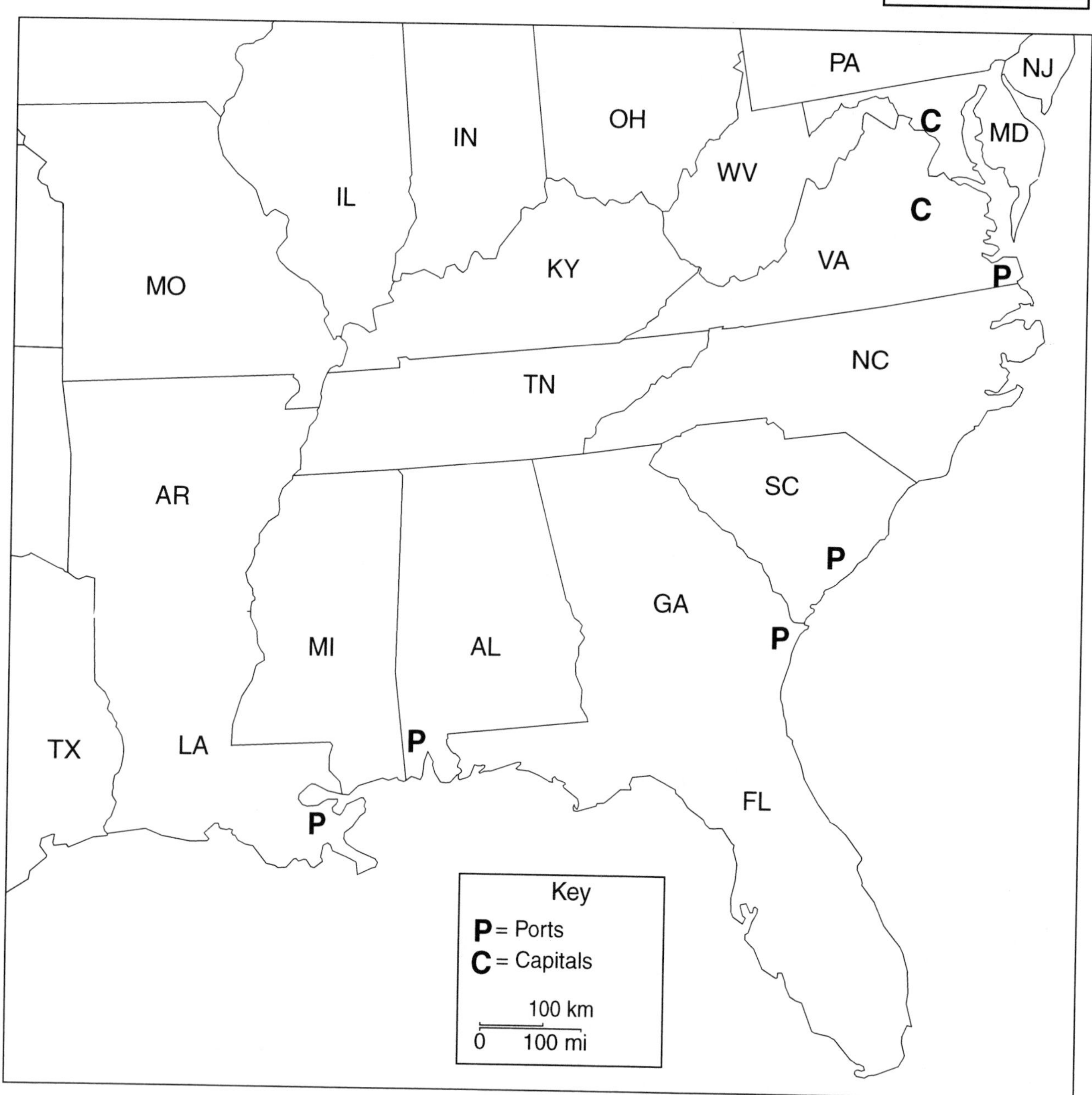
PA
NJ
IN
OH
C
MD
IL
WV
C
MO
KY
VA
P
NC
TN
AR
SC
P
GA
P
MI
AL
TX
LA
P
FL
P
Key
P = Ports
C = Capitals
100 km
0 100 mi

Name ______________________________

Date ______________________________

CIVIL WAR GAME

TOKENS FOR FORCES

UNION FORCES

Start—Army Forces

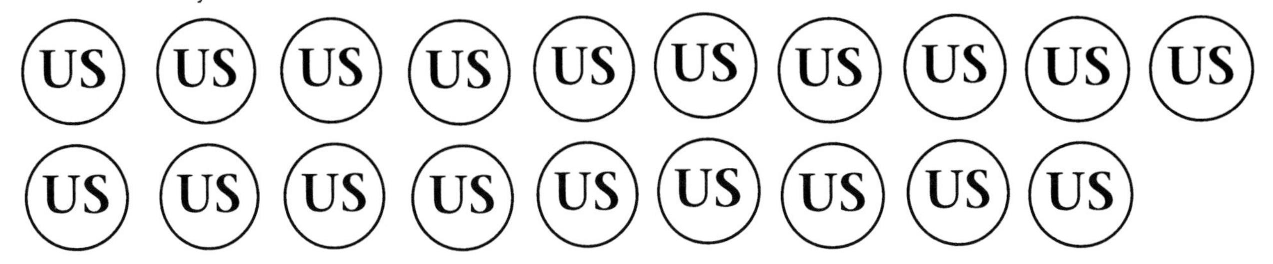

Start—Fleets

Extra Corps

Extra Fleets

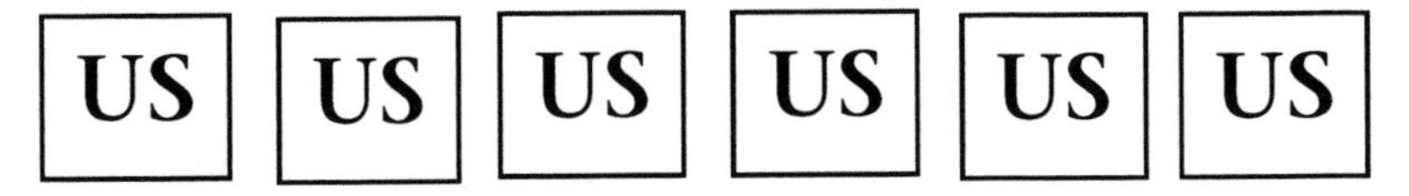

CONFEDERATE FORCES

Start—Army Forces

CSA CSA

Name ______________________

Date ______________________

CIVIL WAR GAME (continued)

Fleets

CSA CSA CSA CSA CSA

Extra Corps

CSA CSA CSA CSA CSA

RANDOM NUMBER DEVICE

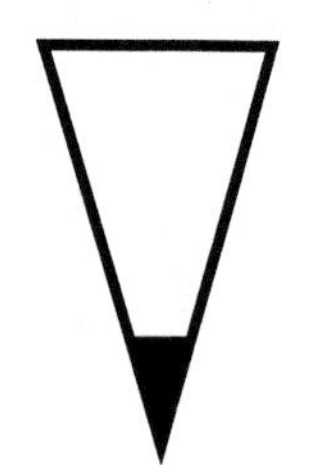

To pick a number, drop triangle on the circle. The sharp point indicates the number.

BRITISH FORCES

Corps

UK UK

Fleets

UK UK UK UK UK

UK UK UK UK UK

FRENCH FORCES

Corps

F F

Fleets

F F F

CHAPTER VIII

THE WILD WEST

For decades Americans watched fictional television images of the West: Marshall Dillon shot bad guys, the Lone Ranger rode again, and Roy Rogers rode happy trails. Those series have ended, but Westerns are still made, and reruns go on forever, so Americans grow up thinking they know a lot about the West.

Return with us now to the days of yesteryear: Let's see what the pages of history really say about Dodge City and other cattle towns. The cattle business in Kansas is historical fact. Cattle in post-Civil War Texas were cheap, and there was a good market for beef in Kansas. Cattlemen could gather a trail herd, drive it to Kansas, and make a handsome profit. There, after selling the cattle, they paid off their trail hands, who usually went on a spree. The rough behavior of the trail hands and lingering hostility between Texans and Yankees gave trail towns a reputation for violence. The following game explores that reputation. It takes a little less than half an hour.

PREPARATION

The only preparation is to make a transparency from the master on page 68. You could also use a chalkboard to present the figures—perhaps writing them out in advance where they could be covered by a map or chart. A bulletin board or poster would work, too, as long as students can't see the numbers before the discussion. I assign advance reading on the cattle business from the text, so some students come to class with background. The class needs no orientation.

A SAMPLE GAME

To establish a factual background, I ask a few questions: Why did people make the cattle drives? What were some of the important trails? What towns were important cattle towns? At this point I am easing into the game.

"Those cattle towns were pretty rough places. How many people do you suppose were killed in Dodge City in a week? I'm just wondering what your estimate would be. How many, Matt?"

Matt hesitates a few seconds, "My guess would be about five a week."

Kitty is always a little more careful. "I would think three a week is more reasonable."

I do not give clues or indicate a favoritism toward either answer. I note each estimate on the chalkboard. "Any other estimates?"

Roy, who is always quick on the trigger, volunteers, "I think 15 a week would be nearer to it."

Dale, brisk and businesslike, supports Roy. "Fifteen looks reasonable to me."

We now have three estimates on the chalkboard. "Are there any other estimates?" If there are, I note them. Then, "First let's get a vote. How many of you think Kitty is right? How many think Matt is right? Those who agree with Roy?" My students raise their hands to indicate their choice. At this point the whole class is committed to a number. "Now let's figure out what the grand total for a season would be. Let's assume an eight-week season. Eight times each of these. Say 24 is a low estimate and 128 is the high." Students are usually nodding agreement.

"We also have to consider that there are other cattle towns. Could we assume about the same for them?" Again, nods of agreement. "Now let's look at a transparency on this subject." I flash the transparency on the screen, and heads stop nodding and start shaking. Students' first reaction is often that the numbers are too low because they are incomplete. I assure them that these are indeed accurate figures. Now is the time to make a few points.

Debriefing

"Why were your guesses so high? What influenced you to think the West was so violent?" My students are usually pretty quick to see the point. For the sake of more exciting stories, the television, movies, and book people have created a legendary West that is much more violent than the real thing.

Indian Fighting

I once offered a slip good for a free A on a quiz or report to any student who could show me documented proof of an Indian attack on a wagon train of settlers. I suspect very few of my students tried to get it, but after some digging, one did locate a case that looked as if it could fit. Notice the limits I used. The army fought Indians, traders fought Indians, and there were attacks on settlements; but attacks on wagon trains of settlers were relatively rare. Perhaps your students will be able to prove otherwise, but if they can't, another Hollywood myth bites the dust.

The Cavalry

Combat was just one of many dangers cavalry soldiers faced. One of the most active and able units in the Regular Army was the Tenth Cavalry Regiment. The Tenth was in a number of campaigns, including the pursuit of Geronimo, action against the Cheyennes, the Spanish-American War, and Pershing's expedition into Mexico chasing Pancho Villa. In those actions they suffered some casualties, but combat was not the only way to die. Two privates died on a long march in 1877 because of lack of water. One soldier had gone 86 hours without water.

You might want to read the above paragraph to your students and ask them what actor they would cast to play a trooper in the Tenth. In fact, the Tenth was made up of African-American troopers. The officers were white. You may want to raise the critical question: Why are so few African Americans shown in movies about the West?

In debriefing after that question, you might want to mention the Pershing expedition into Mexico as an example of the multicultural nature of the West. The force that chased Villa was multiracial, including not just black and white soldiers, but Apache scouts as well. In Mexico, a number of Chinese approached the regiment for protection and returned to the States with them. That's cultural and racial diversity on a scale you don't see in the Hollywood version of the West.

Bibliography

Dykstra, Robert R. *The Cattle Towns*. New York: Atheneum, 1970. (This scholarly history of the Kansas cattle towns is the source of the table on cattle town homicides. It is a serious social, political, and economic history that completely avoids the Hollywood shoot-em-up tradition.)

Clendenen, Clarence C. *Blood on the Border: The United States Army and the Mexican Irregulars*. New York: The MacMillan Company, 1969. (This is mostly a military history, but it's not as lurid as the title suggests. It surveys Mexican-American relations from the 1850's to about 1930.)

Glass, E .L. N., ed. *The History of the Tenth Cavalry, 1866-1921*. Ft. Collins, CO: The Old Army Press, 1972. (This is a reprint of a 1921 history. To read it is to gain a new respect for the soldiers of any race on the frontier.)

CATTLE-TOWN HOMICIDES[1]

Town	1870	1871	1872	1873	1874	1875	1876	1877	Totals
Abilene	2	3	2						7
Ellsworth			1	5	0	0			6
Wichita		1	1	1	1	0	0		4
Dodge City							0?	0	0
Caldwell									0

Town	1878	1879	1880	1881	1882	1883	1884	1885	Totals
Abilene									
Ellsworth									
Wichita									
Dodge City	5	2	1	1	0	3	2	1	15
Caldwell		2	2	3	1	2	2	1	13
Total									45

[1]Robert R. Dykstra, *The Cattle Towns*. New York: Atheneum, 1970, p. 144.

CHAPTER IX

THE RAILROAD GAME

Sometimes a most unlikely topic can be the basis for an interesting game: railroads, for example. Because railroads appear to be sick or dying all over the United States these days, it's a little difficult to imagine a time when railroads were a booming business and had the power to oppress people. Yet in the nineteenth century, farmers complained that railroad rates were ruining them. Railroads tended to set low freight rates for routes between major cities because they were competing for that business. In rural areas, however, railroads had monopolies and could charge "all the traffic will bear;" and high freight rates pushed already depressed farm prices even lower. Farmers' anger and resentment led to the Granger laws and, eventually, to the Interstate Commerce Commission—all of which sounds more than a little tedious and obscure over a century later. How can a teacher keep the class awake while explaining this situation?

The best way I have found is to put students inside the situation and let them work out the problem that railroad management faced, so that they come to feel the problem and live with it. The role playing is easy to set up, and the game takes roughly 20 or 30 minutes.

PREPARATION

In this game it is helpful to have students read about the farmers and their railroad problem before playing the game. Assign them the appropriate material in your textbook. If you have no reading material on this topic, you can introduce the information in a short lecture. You could also just play the game and let them discover the problems. It would probably work, but I have never tried it that way. If you are using the overhead projector, you need to prepare the transparency. If you are using the chalkboard, all you do is take a deep breath and start drawing and talking.

A SAMPLE GAME

Assigning roles is the key to making this game work. I use only about five players. The rest of the class watches and perhaps whispers advice to a nearby player. You need to select people who can understand the problem and react competitively. The long-haul shipper plays a crucial role. It is most important that he or she be a shrewd, calculating type who will select the lowest rates available, thus forcing the presidents of the two railroads to bid against each other for the long-haul traffic—assuming that the presidents of the two railroads will themselves be entrepreneurial enough to compete for the business.

Once I made Diane, a sweet girl without a competitive bone in her body, president of a railroad. She wouldn't cut her long-haul rate, and she kept her short-haul rate low. I nearly panicked as I hinted, "You're being very generous here. Are you sure this is the rate you want?"

Eventually some of the benchwarmers convinced her she should adjust her rates. The role of short-haul customers is not critical. They have no decisions to make. They can ship only on one railroad and are thus at the mercy of that line. Usually after a round of rate setting, I ask, "What line will you ship on?" One of them will reply, "I have no choice." That is a key idea in the game: Farmers were caught in a situation that allowed no choice.

After assigning roles, we start setting rates: "Mercedes, what will your rate be for freight from A to B?"

"Ten dollars."

"What will be your rate from C to B?"

"Five dollars."

I note these rates on the chalkboard and turn to Keisha, the rival line's president, and ask the same questions. The AB rate is fifteen dollars, and the DB rate—this line's short haul—will be twenty dollars.

Now I turn to Micah, my shipper at A, and ask, "Which line will you choose to ship your freight?"

Micah quickly replies, "I'll ship on Mercedes' line for ten dollars."

Then I turn to my two short-haul shippers and ask them, "Which line will you ship on?" They have to agree to ship on the only line that serves them, but one of them is paying considerably more, a point not lost on Mercedes.

This is a good time to total up the amount of freight business each railroad is doing, to spur competition, but usually it is not necessary. When I turn to Mercedes to have her set her rates for the next round, she is ready.

"I'll charge nine dollars for the AB route and thirty dollars for the CB route."

Keisha responds with, "I'll charge eight dollars for the AB route and thirty for my DB route."

At this point I try to start an auction on the long-haul rate. "Mercedes, will you adjust your AB rate? What will you charge?"

"Seven dollars."

Keisha is adjusting quickly, too. "Six dollars."

Typically, my two railroads reach a point where they are charging five dollars or less for the long haul, AB, and up to one hundred fifty dollars for their short routes, CB and DB. These kids are not given to subtle variations!

One recent game had an interesting development. I was just running my rate auction when Keisha called a halt. You could almost see the light bulb hovering over her head as in the cartoons. "Could I talk to Mercedes for a minute?" I let them talk; then Keisha announced, "We'll both charge one hundred fifty dollars for the AB route." My students had just invented a railroad pool. They were fixing rates. Jackpot!

More typically, after about 20 minutes, my students reach a point where long-haul and short-haul rates are markedly different and the class understands why this happens. The short-haul customer, who represents the farmer, is usually a little frustrated.

Debriefing

A few questions drive the point home: Why did the railroad rates turn out as they did? Why were long-haul rates so low? Why were short-haul rates so high? These questions can be directed toward students who played shippers or railroad managers or to observers in the class. It's worthwhile to ask, "How close were our rates to actual historical rates?" Typically, our rates have been much higher. Our rates were just hypothesized numbers and were not expressed per ton or per carload. Real national averages do not differentiate between long haul and short haul, but they do suggest that our rates were unrealistically high. In 1899 transporting a ton of freight generated $1.82 for railroads. In 1900 the revenue was $1.80. Even such apparently low rates could be pretty steep for farmers: In 1899 corn sold on average for $.30 a bushel, wheat for $.59 a bushel, and cotton for $.069 a pound. In 1900 the figures were corn, $.35; wheat, $.62; and cotton, $.0915. Local markets, in turn, were affected by railroad freight rates. A display in the Ames, Iowa, Public Library pictured a shipment of corn that sold for $.11 a bushel in 1897. In June 1997, corn was $2.68 a bushel, wheat was $3.83, and farmers were reported to be doing well.

I ask my students how these rates affected farmers, which breaks down into two questions: How did freight rates lower selling prices for farmers' produce, and how did they raise the prices of items the farmers bought? In Iowa, where I teach, students have a vague idea that farm prices move up and down constantly and that local prices depend on activity at a few large markets—Chicago, for example. The local price for grain is determined by the Chicago market price, minus the cost to ship the grain to Chicago. For example, if the Chicago price is $3.43 and the freight cost is $1.00, the local price would be $2.43. If the freight charge is $1.50, the local price would be $1.93. Thus, high freight charges lower the local market price. Farmers are at the mercy of both market fluctuations and freight rates.

For items that farmers must buy, local merchants can set a fixed price because a town often has only one farm implement dealer: a local monopoly. Even when there are several dealers, they can easily agree to keep prices at a set level, including freight in their costs and adding a profit. Thus, high freight charges lower inflows and raise outflows for farmers. Is it any wonder, then, that farmers at the turn of the last century viewed railroads with a deep hostility?

Now, let's examine those Granger Laws.

Modifications

You can modify the game by breaking the class into five teams for the student roles, which allows more class participation. Each team then needs one spokesperson. Usually the group selects the most assertive member as their spokesperson, but you could choose the students for key roles and assign others to be advisors. You are the expert on your class.

There is a comparable later historical development. In the 1980's deregulation became the new trend, and airlines were allowed to set their own prices. Soon the airline tickets were following a pattern very similar to the old-time railroad shipping charges. Prices on highly competitive routes between major cities dropped; those on less-competitive routes to small or medium-sized cities, like Des Moines, Iowa, were much higher. You might ask one or more students to call a travel agent and price air travel to New York, to Los Angeles, and to Des Moines (or any other smaller city). How do prices compare? How do the distances compare? Who wins and who loses when prices are set by airlines (or railroads) without government regulation? In the nineteenth century, farmers were able to get government regulation of railroad rates. Why did we deregulate airline rates in the 1980's?

The Railroad Game

The Rules

These rules are only for the teacher, who is acting as control. (Not handing out rules to students leaves you free to improvise.) The game usually moves forward in the following steps:

1. Either sketch two imaginary railroads on the chalkboard or project them on the overhead using the transparency master supplied for the arrangement.
2. Assign students' roles:
   (a) One student president of the ACB line
   (b) One student president of the ADB line
   (c) At least one student shipper at A, with freight to be shipped to B
   (d) At least one student shipper at C, with freight to be shipped to B
   (e) At least one student shipper at D, with freight to be shipped to B
3. Ask each president to announce his or her railroad charges to transport freight for the long haul from A to B. Note the rate on the chalkboard or overhead.
4. Ask each president to announce his or her railroad charges to transport freight for the short haul from C to B or D to B.
5. Ask the long-haul shipper which line he or she will ship on.
6. Ask the short-haul shippers which line they will ship on.
7. Repeat steps 3, 4, 5, and 6 as often as needed. Try to get the two companies to compete. You will probably find that you get an underbidding auction going on long-haul rates, while short-haul rates skyrocket.

Name ____________________

Date ____________________

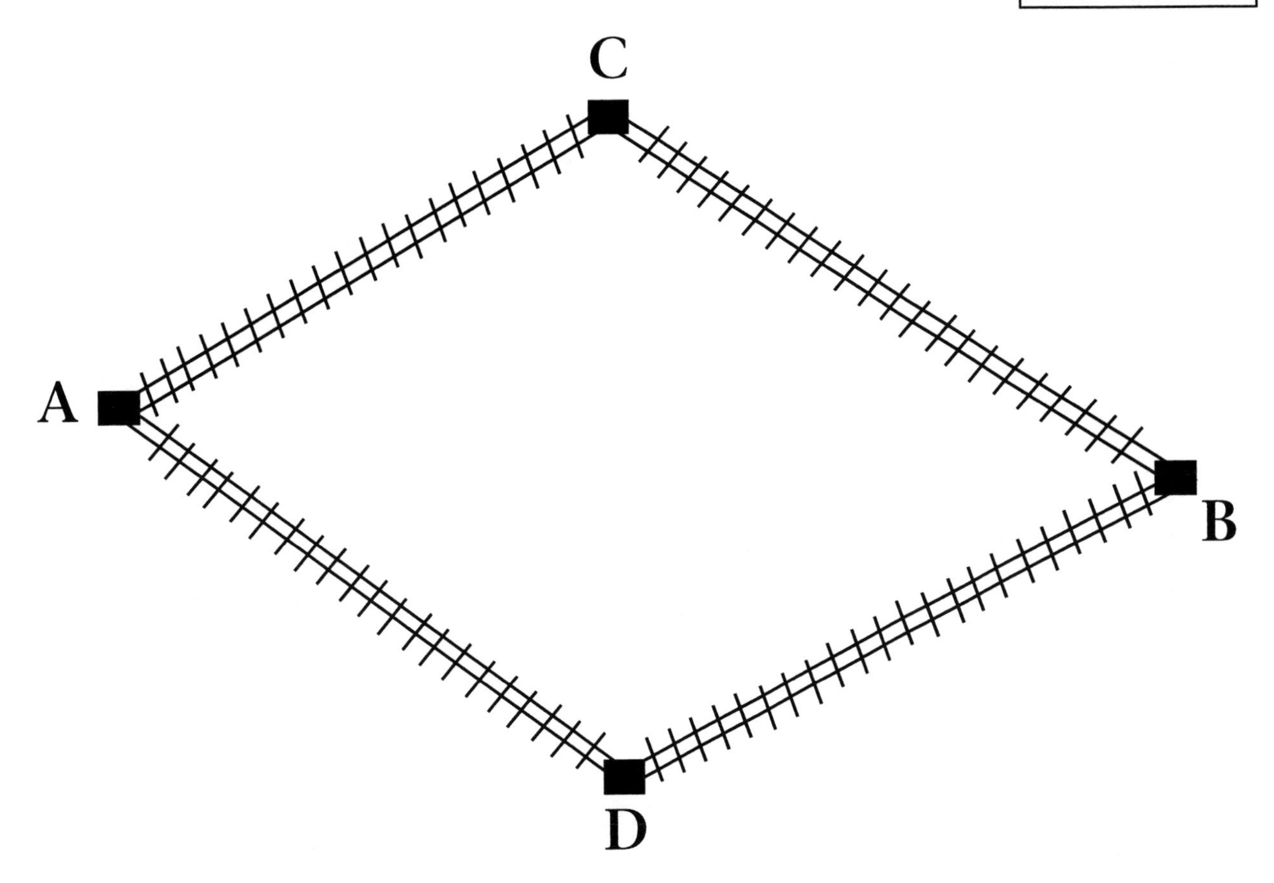

ACB RAILROAD RATES

A to B

C to B

ADB RAILROAD RATES

A to B

D to B

CHAPTER X

1898

My students are fond of hypothetical questions about past events and behavior: "Why didn't they . . .?" Usually the alternatives they suggest are anachronisms. In the following game, you give students the actual information that was available to a president; then ask them to make decisions. The president in this case is McKinley, and the policy decision revolves around U.S. relations with Spain in 1898. If you play this game before students have done any reading on the topic, the effect is to minimize hindsight as they sort out relevant issues and deal with conflicting pressures in the president's place. As I use it, the game explores the specific pressures related to the Spanish-American War. It can also be used to explore the way presidential decisions are made. Best of all, it lets you ask your students, "Why didn't you ...?" The game takes about half an hour.

PREPARATION

To reiterate, it is probably best to run the game before studying the Spanish-American War, to avoid bias in making decisions.

A day or so before the game, you should make copies of the handouts (see pages 67–73), one full set of handouts for each five students in your classes. It is also a good idea to separate the handouts you distribute at the start of the game from those you distribute as the game progresses. Be sure you have the reaction sheets organized so you can hand them to students once the president has made a decision.

ORIENTING THE CLASS

On the day of the game, simply call the class to order and organize them. First, organize groups of five students, a president and four advisers. For a random distribution of talent, have them count off: one, two, three, four, five. Then all the "ones" are president, the "twos" are political advisers, and so on. You could also handpick presidents from the more able students in your class (a bit unrealistic in this case: President McKinley was a pleasant man, but nobody ever accused him of being exceptionally intelligent). Once the groups are formed, you need to explain the game. It runs something like this.

"Today we are going to be playing a little game. In a few minutes, I'll be handing you a sheet of paper which will tell you your role in the game. One person in each group will be President McKinley, who will lead the group. When you get the sheet of paper, read it carefully. Each of you has important information that others do not have.

"Then, the president should gather you into a group for discussion. After the discussion the president needs to make a decision. When I get that decision, I'll probably have a set of reactions to it. Any questions? No, Brad, you will not be graded on getting a right answer. We will start now."

A SAMPLE GAME

Now, move quickly to pass out the sheets of information. At first, the students may be puzzled, but just reassure them that their information will all fit together in the discussion. You may want them to sit, read, and think about their roles for a few minutes. Then the groups should gather and begin discussing the problem.

Depending on the groups, you may have to quiet them down or urge them to talk.

Let's follow some sample cases through the process. Brenda is an efficient president who briskly gets her group to work. After about 10 minutes, she makes a decision and presents you with an action sheet. She has checked Option 1, *Ask Congress to declare war on Spain*. You then return to her group and share the reactions keyed to that choice. This new set of information calls for another decision. The president and her advisers must now choose a naval strategy for the war and write a brief statement of what they decide. Brenda decides to scatter parts of the fleet in each major city on the East Coast. Once she has handed you the statement, her group has finished the game.

Cathy gets her group together but seems more focused on an upcoming party Friday night than on being president. After a reminder, she dutifully plunges into a discussion of the assigned problem and soon presents you with an action sheet. She has checked Option 2, *Instruct the Secretary of State to continue diplomatic efforts to solve the problem*. You respond by presenting the group with the reactions for Option 2. (By now you have action sheets coming in from most of the presidents. Aren't you glad you sorted those reaction sheets so you can grab them fast?) You tell Cathy to respond on the same action sheet she used last time. After a brief consultation with her group, she picks the first option, *war*. You present the group with a reaction sheet for that option. Cathy now has to decide on a naval strategy. Once she has done that, her group has finished the game.

Juan has a creative mind. As president, he quickly comes to a decision, hands you an action sheet, and then his group takes up the topic of motorcycles. However, you are too quick for him. His group has chosen Option 3, *Recognize the insurrections in Cuba and allow shipments of weapons to them*. You promptly present them with the reaction sheet for Option 3, which calls for him to formulate a policy to deal with several groups of *insurrectos* and an irate Spanish government—tasks that should challenge his creativity. Once he has handed you his solution to these problems, his group has finished the game.

During the game you have to circulate to keep the groups on track. Some work faster than others. When to call a halt is a matter of judgment: too late and concentration is gone, too early and the process is unfinished. I try to stop the game when all groups have made their first decision and one or two groups have finished.

Debriefing

My debriefing assumes that the exercise is being used to explore specific pressures brought by the Spanish-American War. If you are using it to explore the general nature of presidential decision making, you may want to modify some questions, although some work for both approaches: How much did public opinion sway your decision? Was one adviser particularly persuasive? Why? Did reaction in news media influence you? Did you consider that we might lose the war? Did the weakness of the army cause you any worry? How does the present military strength of the United States compare with our strength in 1898? Would public opinion today reflect the same attitude toward such a war?

There are a few items you might want to weave into the discussion. The handouts students received are based on historical information. The people credited with writing them are people who would have written them for President McKinley. The numbers of troops in the army are actual statistics. The judgments made in the handouts were those really made at the time. The handouts are reconstructions, however, not actual documents.

There is historical support for each reaction. It is easiest to tell what the reaction was to the first choice, war on Spain: This was the choice McKinley actually made, so the historic record is there. Reactions to the second choice are a little more conjectural, but the president had dealt with earlier incidents by diplomatic efforts, and public reaction had been unfavorable. Reactions to the third choice are the most speculative. They are based on two assumptions. It seems almost certain that the Spanish government would protest. As a matter of record, the *insurrectos* in Cuba were not a single organization. It is quite probable that more than one group would apply for recognition. The choice of the number three was purely arbitrary.

If any group actually made it to the strategic decision about how to fight a war, they faced a mess. Each East Coast city wanted naval ships to guard it. The navy did not want to scatter its forces against the Spanish fleet sailing from Spain to Cuba. The *Oregon*, one of our best ships, was on the West Coast where it was useless. McKinley and his advisors handled it neatly. The *Oregon* was ordered to sail from the West Coast to the East Coast, going around South America. There was no Panama Canal yet. While a few obsolete ships guarded the cities, the navy formed a strong fleet and eventually cornered and sank the Spanish fleet.

The description above is obviously very simplified and full of generalizations. All generalizations are false, including this one, so brighter students might challenge them. This is great, because it gives you a chance to point out that the game is a simulation. Now you have a good opportunity to have students do research on the real thing. You can assign either oral or written reports and even ask them to rewrite and correct handouts based on their research. The following bibliography has a selected list of sources.

Bibliography

Freidel, Frank. *That Splendid Little War.* Boston: Little, Brown & Co., 1958. (An introduction to the war featuring many pictures and a bibliography. Out of print, but worth an interlibrary loan request.)

Werstein, Irving. *1898: The Story of the Spanish-American War and the Philippine Insurrection Told with Pictures*. Totowa, NJ: Cooper Square, 1966. (Packed with contemporary photos, drawings, and facts.)

1898

The Rules

The teacher acts as control and directs the class through the game.
(You should not distribute these rules to students.)

1. The class is divided into groups of five, in which each student is assigned one of these roles: President McKinley, Secretary of the Navy John D. Long, First Assistant Secretary of State William R. Day, Secretary of War Russell A. Alger, and a political adviser.
2. Cabinet officers and advisers receive a handout explaining the background on the situation. Each president receives a set of directions for what to do.
3. The groups exchange information and make a decision. Once the decision is made, the president makes a check on the action sheet indicating the decision.
4. The president then brings the action sheet to control, who in turn gives the group the appropriate reactions and a new action sheet. The general rule is that a group is given a set of reactions with the number matching its action: action 1 gets reaction 1, action 2 gets reaction 2, and so forth.
5. The group then studies the reaction and notes another decision on the action sheet. (If they choose action 1, the decision will have to be written in a sentence or two. If they choose either 2 or 3, they can use the same action sheet as in step 3.)
6. Control decides how long the process runs and directs the debriefing. The groups all should go through steps 3, 4, and 5 at least once, perhaps more, depending on decisions they make.

Name ______________________________

Date ______________________________

PRESIDENT MCKINLEY

As president of the United States, you are facing a serious crisis. The year is 1898. For several years the United States and Spain have argued about Cuba. Now the argument has reached the point where you must make a decision. You have four advisers who each have information and ideas that can help you decide. In deciding, you may want to consider the following questions. Is a war with Spain necessary? Could we solve the problem by diplomacy? Could we win a war with Spain? Would the public support a war with Spain? How will the action you take affect your political future? When you have discussed matters thoroughly with your advisers, check the response matching the decision you have made, and turn the action sheet in to the teacher.

ACTION SHEET

____ 1. Ask Congress to declare war on Spain.

____ 2. Instruct the Secretary of State to continue diplomatic efforts to solve the problem.

____ 3. Recognize the *insurrectos* in Cuba and allow shipments of weapons to them.

Name ______________________

Date ______________________

SECRETARY OF THE NAVY JOHN D. LONG

In this report I shall discuss the explosion on the *Maine.* I shall then describe the strength of the U.S. Navy. On February 15, 1898, the battleship *Maine* was sunk. There was a tremendous explosion. At the time, the *Maine* was peacefully riding at anchor in Havana harbor. Over 250 officers and men were killed. The navy has investigated the explosion. Our conclusion is that the explosion was from a mine. The plates of the ship's hull were bent inward. The navy does not know who placed the mine.

The U.S. Navy compares favorably with the Spanish Navy. In guns in the main battery of our ships, we outnumber Spain 244 to 160. In armored tonnage built and building, the United States has 152,000 tons opposed to Spain's 90,000 tons. In general, our ships tend to be newer than the Spanish ships. The loss of the *Maine* does, of course, weaken our navy, but we have other battleships.

The Asiatic Squadron of the U.S. Navy is now located at Hong Kong. Commodore Dewey has them ready for action. They are in an ideal position to attack the Philippine Islands. These islands are a Spanish colony. The *Oregon,* a first-class battleship, is unfortunately not in a good location. It has just come out of drydock in Bremerton, Washington. If it is to see action near Cuba, it must steam around South America to join the fleet. In general, the U.S. Navy is fit and ready for action.

Name ______________________________

Date ______________________________

FIRST ASSISTANT SECRETARY OF STATE WILLIAM R. DAY

In this report I shall discuss the record of our troubles with Spain and the latest diplomatic developments. The Cubans have been in rebellion since 1895. Since then, the island of Cuba has been the scene of death and destruction. Spain has resorted to rounding up all the population and placing them in reconcentration camps. Spanish authorities have also, as a matter of policy, destroyed any property in Cuba that could be used by the rebels. Unfortunately, these policies have also led to imprisonment of American citizens and destruction of American property. In addition, the American public has been offended by the bloody methods of the Spanish. We have repeatedly protested the treatment of U.S. citizens and property. The Spanish have protested to us about weapons being smuggled from the United States to Cuba. We have tried to stop this but must admit that we are not completely successful.

Recent events have been promising. General Weyler, the bloodthirsty Spanish general, has been replaced. We take this as an indication of moderation by the Spaniards. We have been pressuring the Spaniards to declare Cuba independent. Our minister in Spain reports that Spain will release the prisoners in the reconcentration camps and declare an armistice with the rebels.

Name ______________________

Date ______________________

SECRETARY OF WAR RUSSELL A. ALGER

In this report I shall discuss the current condition of the U.S. Army. At this time the U.S. Army is too small to fight a war with Spain. Our most recent fights have been with Indians. We presently have 10 regiments of cavalry, 7 of artillery, and 25 of infantry. Our total strength is about 2,000 officers and 28,000 enlisted men. The Spanish have about 200,000 men in Cuba. They are armed with modern Mauser rifles. Our forces are equipped with the new Krag-Jorgensen rifles. We have no extra weapons for new recruits. In the event of a sudden expansion, we would have to issue the older Springfield rifles. Due to the lack of a general staff, we have no detailed plans for handling an expanded force.

While the above may not look very promising, I would point out the following. All over the country our recruiting stations have been swamped with volunteers. Willing men are available. We have faced similar problems in raising, equipping, and training an army before. During the Civil War we were able to function adequately in a similar situation. If war is declared, the army will go about its mission with spirit and enthusiasm.

Name ______________________

Date ______________________

POLITICAL ADVISER'S REPORT

In this report I shall discuss the political situation and public opinion. The Republican Party needs vigorous leadership from you, Mr. President. This is a congressional election year. The newspapers are screaming for a war. They have reported in gory detail the atrocities of the Spanish armies in Cuba. The public is outraged over an incident in which Spanish officials reportedly searched a lady traveler. Newspapers have obtained and published a letter by DeLome (the Spanish minister in Washington) calling you a "would-be politician." Their editors are now screaming, "Remember the *Maine*!" They assume the Spanish government planted the mine that blew up the *Maine*. Generally, the public believes these press reports, and a real war fever is sweeping the country. In such a situation, the Democrats could destroy us in the next election if you don't do something vigorous soon. Even within the Republican Party, voices can be heard calling for war. A few of the more solid, prosperous citizens in the United States are opposed to war, but they are in the minority.

Name ______________________________

Date ______________________________

I. REACTIONS

POLITICAL

The country is swept up with a great wave of enthusiasm. Your popularity is soaring.

NAVY

Commodore Dewey reports that he has destroyed the Spanish fleet in Manila Bay—a great victory!

On the East Coast of the U.S., we are swamped by requests for protection from the Spanish fleet. Our naval planners think we should keep our fleet together and not scatter it along the coast. What should we do about the *Oregon*? Please advise.

ARMY

Volunteers are swarming in. We are organizing as rapidly as possible to prepare them to go to Cuba.

DIPLOMATIC

The European governments are registering disapproval of our action. So far, there is no sign that they will help Spain.

Name ______________________________

Date ______________________________

2. Reactions

Political

The newspapers are outraged. The Democrats are jubilant and getting ready to ride to victory on the wave of public reaction. You are being labeled a coward.

Diplomatic

The Spanish report they are still working on an armistice. They are not yet ready to grant independence. Please advise.

3. Reactions

Political

Public opinion is split. Many see your decision as too little too late. Others support it. The Democrats are preparing to attack you in the coming campaign.

Diplomatic

The Spanish government is protesting your action in the strongest terms. They cannot allow you to interfere in their internal affairs. Three groups have applied to be recognized as the Revolutionary Government of Cuba. Please advise.

CHAPTER XI

THE STOCK MARKET GAME: A SIMULATION OF THE GREAT CRASH OF 1929

"Mr. Scott, last night I dreamed you crashed the stock market." It's not often that students dream about their American history class, at least about the subject matter. Not many activities have students stopping me in the halls to get information. "Did stock prices go up today?" "I've been gone. What happened to RCA?" This stock market game generates a lot of interest. It also shows students the actual prices of stocks in the 1920's. They get the long upward climb, the peak, and the crash.

It's a fairly long-running game, and students have to do a little arithmetic. In my classes the game takes up part of the class period for ten days. On the first day, it uses about half an hour. During the next 8 days, buying and selling stocks takes 10 to 15 minutes. The tenth and final day includes debriefing and usually requires a full class period. The students must be able to calculate the price of a stock purchase. The cash formula is simple: Price per share times number of shares equals total price of the shares purchased. For margin buying, students also have to be able to handle percentages. They pay 10 percent down and borrow the remaining 90 percent. Usually students master this with a little help from their calculators.

It's an easy game to run. The teacher simply displays stock prices each day, helps those who are confused, and offers a little patter to add flavor.

PREPARATION

Make transparencies of all the masters on pages 80–88, and be sure to have an overhead projector available for the next two weeks. If this is not possible, write the stock prices on the chalkboard or read them aloud to the classes. You could also use a bulletin board to display stock prices, with removable cards to change prices each day. Any device that lets you post stock prices one day at a time is satisfactory. You will also need 10 copies of the student record form for each student. Do not hand out all 10 at once because that will telegraph the end of the game. In my class I hand out a sheet with 4 copies of the form, 2 per side three times during the game, on days 1, 5, and 9. That makes it a little harder to figure when the end of the game is coming.

ORIENTING THE CLASS

This is a little like a magic show. It's what you conceal that makes it interesting. I am very open about the rules and the risks involved, but very secretive about some other things. I start by flashing the stock prices on the screen. "Today we are going to start playing a game about the stock market. You will start with $1,000. You may use it all to buy stock. You may buy any combination of shares as long as you don't spend over $1,000. You may buy stock on margin, that is, you may pay 10 percent down and borrow the rest. Buying on margin can make you a lot of money, but it can also be very risky. If a stock you buy on margin goes up 10 percent, you have doubled your money. If a stock you buy on margin goes down, you have lost all your money. Note that if the price drops 10 percent or more, you are forced to sell the stock and repay the margin loan.

"Remember, the stock market can go up and it can go down. Are there any questions?"

My students consider the problem with a skeptical eye. At least one student in each group mutters, "He's going to crash it." I do not respond.

Others try to beat the game. "How long will this game last?" Toh wants to know.

I reply, "I do not answer that question."

Grant asks, "Are these stock prices real? How do we decide what to buy?"

I tell him, "There is a real basis for these prices. At this time I will not tell you what it is. After the game is over, we can discuss it. I can tell you, however, that all companies listed were real. You are pretty much in the dark now about what stock to buy. As the game progresses, you will see each stock develop a history that will help."

Tina has a practical question. "Do I have to buy stock?"

"No, you don't have to buy stock. You may keep your money in the bank. If, at a later stage you want to buy stock, you can do it then, but you must do it at those prices. You will never see these prices again."

Jim has a problem. "I have to be gone tomorrow. How will I buy stock tomorrow?"

"You can't buy or sell stock when you are not here. It's just like the problem of a real investor who is on vacation and is not able to contact a broker. When you return, you have just missed a few days. You still own the stock you bought before you left."

At this point I hand out the sheet designed as the students' record of stock transactions. In my version each student gets four copies of the sheet, enough for four days. Other copies are handed out later. I then walk them through two sample transactions. "Let's assume you want to buy General Motors stock. It has a price of $10. You would write GM on the form where it says "Name of stock bought," price, $10; number of shares, 100. You have paid $1,000 cash; you note that and put zero in the cash-held box. You have just invested all your money.

"If you want to buy GM on margin, the process is a little different. The stock and price are the same, but now you can buy 1,000 shares. You pay $1,000 and borrow $9,000 on margin. You are holding no cash, and you owe $9,000 on margin. When you sell the stock, you will have to repay that $9,000. Please note that I am not trying to push GM stock. I have chosen that stock because it is at a price that makes it easy to figure. Now go ahead and make your decisions for the day." I don't say, ". . . make your stock purchases." Pushing people into the game just irritates them. Usually as the game progresses, even reluctant students decide to get in on the action.

Now that students are figuring their investments, there are a few further questions. The most common is, "What's $^7/_8$?" I explain that the fractions are fractions of a dollar. They know that $^1/_2$ is $.50 and $^1/_4$ is $.25. They can figure out that $^1/_8$ is $.125 and that $^7/_8$ is $.875. This system of fractions is now obsolete, but it was in use in the 1920's.

It usually takes about 10 minutes for the students to get their first stock purchase figured out. I circulate among them, dealing with questions about the forms and generally seeing how they are doing. After a reasonable time, I shut off the overhead projector and tell them to put the stock market material away. I remind them to bring the sheets and perhaps a calculator to the next class. We are now off and running on the stock market game.

A Sample Game

In general I like to add a little flavor to the game by making pronouncements. These are either old jokes about the stock market or quotations from the period. These three are my favorites. "The secret of making money on the stock market is simple: buy low, sell high." "I'll tell you what the stock market will do in the future. It will fluctuate." Those will work anytime, but I always save "Business conditions are fundamentally sound" for the round before the crash.

Some rounds of the game raise unique problems. The second day is a little demanding because for the first time, students can also sell stock. I list their options: "You can decide just to hold what you own, but remember, you haven't actually made any money until you sell your stock and get the money. If you bought the stock on margin, remember to pay back the margin loan when you sell. If you decide to sell, you can buy other stock or just hold on to your profits." This particular day the game tends to take about 15 minutes because some students need to go get their sheets and others need help with arithmetic. Generally, they are making money, and some of the cautious ones who held out on the first round decide to buy a few shares. Some of the more risk-tolerant students plunge into margin buying, investing heavily in RCA. There is nothing like a rising market to generate enthusiasm.

The third day is sobering for the RCA speculators. RCA is down to $4^3/_4$ from $6^1/_4$. Those who were heavily committed on margin have not only lost all their money but are in debt, which could mean that they are out of the game. Rather than see that happen, I offer them loans of $500. Some of them go right back into RCA on margin! I encourage them with a stock tip: "RCA looks good." Students tend to be suspicious of my tips, so they do not flock to RCA.

On the fourth day, they wish they had. RCA has jumped from $4^3/_4$ to $66^7/_8$. Those who invested their borrowed $500 in RCA on margin are now able to pay back the loan and gloat over a sizable profit. At this stage I start asking, "Who here has made the most money? How did you do it?" Some budding financial wizard explains how he bought RCA when it was at $2^1/_2$. Some cautious souls have decided they have made enough. They sell their stock and decide to sit out the rest of the game.

The fifth day is profitable for all concerned, but RCA owners have a new complication: RCA has split. This means that for every 5 shares they had yesterday, they own 6 today. If they had 25, they now own 30, and so on. (Fractions of a share should be dropped.)

The sixth day introduces more complications. American Can has gone through a 4-for-1 split and a 50 percent stock bonus. Each old share is now replaced by 6 new shares, and the result is a modest increase in value. The General Electric situation is similar: 4-for-1 split, each old share replaced by 4 new shares, and a modest increase in value.

The seventh day usually produces a lot of happy investors. All stocks are up except for General Motors and U.S. Steel.

The eighth day is euphoric. All stocks rise. RCA skyrockets, more than quadrupling in value. This is a good day to poll the class again about how much money they are making and how they are doing it. By now, some students will have emerged as stock market wizards, having made millions of dollars. Some will be managing investments for other students. They may be simply saying, "Buy RCA," but they can also form an organization and pool money to be managed by one student. As control, I am neutral on the practice. By now the markets are a routine of the class, taking about 10 minutes of class time before we discuss politics of the 1920's or hear a report on Prohibition.

The ninth day everybody makes money except for the cautious few who sold out earlier. The student investors have mastered the market. They quickly calculate that today's RCA stock split yields a tidy profit. Euphoria is once again in the air. I intone, "Business conditions are fundamentally sound."

The tenth day is the crash. A lot of investors who thought they were millionaires find they owe massive debts on margin loans. At this point I instruct the students to figure out their worth, and we poll the class. Some have made millions. The record debt is around $4 million. Now it's time to talk about the game and how it relates to the stock market of the 1920's.

Debriefing

Now is the time to let the students in on things concealed before. The process involves a lot of teacher talking. In general, the stock market prices were real prices of stocks in the 1920's. From days 1 through 7, the prices were year-end prices. After that the prices reflect the peak of the boom and the sudden drop of the crash. For details, see the time chart transparency.

It's interesting to note how stock prices reflect a company's performance. RCA starts out very low. At the beginning of the game, radio is a toy for a few scientifically minded tinkerers. During the 1920's it expanded into a major entertainment industry. In 1921, 32 stations were licensed to broadcast; by 1922 more than 600 were. RCA produced broadcast equipment and the receivers that appeared in American living rooms. Their revenue expanded from $2 million in 1919 to $182 million in 1929, and RCA stock prices reflected that development. The last stages before the crash also show a strong surge of speculation that pushed prices to record heights.

General Electric was also involved in radio. In fact, at one time GE owned RCA but was forced to sell it due to antitrust laws. In the 1920's General Electric manufactured electric appliances—electric refrigerators, vacuum cleaners, and washing machines. The company became the leading manufacturer of appliances and has remained so into the 1990's.

American Telephone and Telegraph had a monopoly on the telephone business. The monopoly included manufacturing equipment and providing long-distance and local service. It kept these monopolies until 1984, when it was broken up in an antitrust action.

During the 1920's, General Motors became the biggest automobile manufacturer in the United States. At first, Ford Motor Company led the way. In the early years, the American auto industry had dozens of brand names. General Motors bought them up. Among the purchases were Oldsmobile, Oakland, Cadillac, Rapid, Welch, Ewing, Elmore, Mason, Republic, Little, and Chevrolet. General Motors also bought other businesses, including Frigidaire, Dayton-Wright Airplane, Dayton Engineering Laboratories (DELCO), and Hyatt Roller Bearing. The DELCO company held the patent on the self-starting ignition system. General Motors began to sell cars designed for different levels of income. They also came out with new models each year. In the 1920's the auto industry grew explosively, and General Motors led the way.

U.S. Steel became America's first billion-dollar corporation in 1901. It produced roughly two-thirds of the steel in the United States.

American Can manufactured cans for canned foods, an expanding market in the 1920's. All of these companies were prospering in the 1920's, and their stock prices reflect that growth and the expectation that it would go on forever.

The margin rule is another feature of the game that matches reality. People really were buying stock on a 10 percent margin—a risky business indeed! Some other speculative devices were in use, but they were too complicated for the game.

One way that this game simplifies reality is to ignore brokerage fees and interest. Brokerage fees are based on a sliding scale. The less you buy, the larger the broker's percentage. In the early stages of the game, when investors had only $1000, brokerage fees would have been quite discouraging, and interest rates would have added another layer of complexity. They would also have given away the time scale if students paid a year's interest each round.

In the real 1920's stock market, the economic growth, investor confidence, and margin transactions reinforced each other to push prices high. This raises a tricky question. How is the stock market related to the health of the economy? To what extent is the stock market simply a market responding to its own pressures, or does it reflect the whole economy? Are stock market crashes or long bull markets simply caused by the similar swings in the American economy? Or, does stock market movement predict the future of the American economy? Could a crash cause a depression? If so, how can this happen? These are difficult questions. We know that since 1929 there have been stock market downturns that were followed by recessions. In some cases the two came together, but in others the market and the economy did not move in the same direction. Was there something about the economy of the 1920's that made a crash particularly damaging? Questions like these are worth exploring with your students.

The More Recent Stock Market

If you would like to study the current stock market, there is a simple game that I have used for years. At the start of the semester, I do a short introduction to the stock market. I use a film to explain what the market is and what common stock is. Then I give my students a week to invest $10,000 in stock. They turn in a paper telling me what stocks they bought and why. They have to include prices and number of shares. I do not include brokerage fees or anything beyond prices. I grade the papers and return them. If a student wants to trade stock during the semester, he or she must have some other student sign a paper as witness that the trade was made at a certain date. At the end of the semester, students write another paper reporting how their stocks did, why they think prices went up or down, and what they conclude about their earlier methods of stock selection. If they made money, they think they are pretty good. If they lost, they think they had bad luck.

Bibliography

Allen, Frederick Lewis. *Only Yesterday.* New York: Harper & Row, 1959. (This is the most interesting history of the 1920's I have seen. It contains an explanation of the crash and is the source of some of my stock prices. High school students like it.)

Galbraith, John K. *The Great Crash, Nineteen Twenty-Nine.* Boston: Houghton Mifflin, 1979. (An economist analyzes the crash and its impact on the economy.)

Sobel, Robert. *The Great Bull Market: Wall Street in the 1920's.* New York: W.W. Norton & Company, 1968. (This is the source of many of the stock market prices used in the Stock Market Game. Sobel does not think the stock market crash was as disastrous as we have been led to believe. His use of sample prices is interesting.)

Name ______________________________
Date ______________________________

THE STOCK MARKET GAME

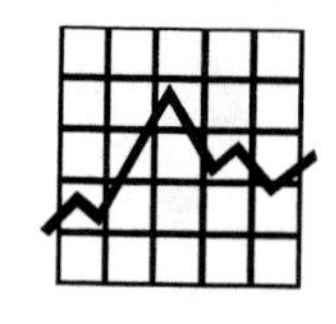

THE RULES

1. Each player will start with $1,000.
2. You decide how much money to invest and how to invest it.
3. You may buy stock on a cash basis. In that case, you figure the cost of the stock and pay the amount in cash. To figure the cost of the stock, multiply price by number of shares.
4. You may buy stock on margin. In that case, you figure the cost of the stock, pay 10 percent margin, and borrow the remaining 90 percent. When you sell the stock, you have to pay back the loan. If the price of the stock goes down 10 percent or more, you must sell the stock.
5. You buy and sell at the prices listed for the current day—that is, on day 1 you buy at day 1 prices; on day 2 you buy or sell at day 2 prices.

Name ______________________

Date ______________________

DAY 1

AMERICAN CAN	34 3/4
AMERICAN TELEPHONE & TELEGRAPH	114 7/8
GENERAL ELECTRIC	139 1/4
GENERAL MOTORS	10
UNITED STATES STEEL	84 1/4
RADIO CORPORATION OF AMERICA	2 1/2

DAY 2

AMERICAN CAN	73 1/4
AMERICAN TELEPHONE & TELEGRAPH	123
GENERAL ELECTRIC	182 1/2
GENERAL MOTORS	14 3/4
UNITED STATES STEEL	106 7/8
RADIO CORPORATION OF AMERICA	6 1/4

Name ______________________________

Date ______________________________

DAY 3

AMERICAN CAN	$104^1/_4$
AMERICAN TELEPHONE & TELEGRAPH	$125^3/_8$
GENERAL ELECTRIC	$196^1/_2$
GENERAL MOTORS	15
UNITED STATES STEEL	$94^1/_2$
RADIO CORPORATION OF AMERICA	$4^3/_4$

DAY 4

AMERICAN CAN	160
AMERICAN TELEPHONE & TELEGRAPH	$130^1/_2$
GENERAL ELECTRIC	320
GENERAL MOTORS	$65^3/_8$
UNITED STATES STEEL	$122^5/_8$
RADIO CORPORATION OF AMERICA	$66^7/_8$

DAY 5

AMERICAN CAN	292 1/2
AMERICAN TELEPHONE & TELEGRAPH	142 5/8
GENERAL ELECTRIC	326
GENERAL MOTORS	117 1/2
UNITED STATES STEEL	126
RADIO CORPORATION OF AMERICA	77 7/8*

....

DAY 6

AMERICAN CAN	49*
AMERICAN TELEPHONE & TELEGRAPH	149 7/8
GENERAL ELECTRIC	83 7/8**
GENERAL MOTORS	153 3/4
UNITED STATES STEEL	157 3/8
RADIO CORPORATION OF AMERICA	61 5/8

* 4-for-1 split plus 2 shares stock dividend. One share of old stock becomes 6 shares.

** 4-for-1 split. One old share is now 4.

Name ______________________________

Date ______________________________

DAY 7

AMERICAN CAN	75
AMERICAN TELEPHONE & TELEGRAPH	178 5/8
GENERAL ELECTRIC	135 3/4
GENERAL MOTORS	138
UNITED STATES STEEL	151 7/8
RADIO CORPORATION OF AMERICA	101

DAY 8

AMERICAN CAN	110 3/8
AMERICAN TELEPHONE & TELEGRAPH	193
GENERAL ELECTRIC	221 1/2
GENERAL MOTORS	203 3/4
UNITED STATES STEEL	151 1/4
RADIO CORPORATION OF AMERICA	420

Name ______________________________

Date ______________________________

DAY 9

AMERICAN CAN	$181^7/_8$
AMERICAN TELEPHONE & TELEGRAPH	$335^5/_8$
GENERAL ELECTRIC	$396^1/_4$
GENERAL MOTORS	$181^7/_8$
UNITED STATES STEEL	$279^1/_8$
RADIO CORPORATION OF AMERICA	101*

* 5-for-1 stock split

DAY 10

AMERICAN CAN	86
AMERICAN TELEPHONE & TELEGRAPH	$197^1/_4$
GENERAL ELECTRIC	$168^1/_8$
GENERAL MOTORS	36
UNITED STATES STEEL	150
RADIO CORPORATION OF AMERICA	28

Name ______________________________

Date ______________________________

TIME CHART

GAME DAY	REAL TIME
1	Close of year 1921*
2	Close of year 1922*
3	Close of year 1923*
4	Close of year 1924*
5	Close of year 1925*
6	Close of year 1926*
7	Close of year 1927*
8	Close of year 1928*
9	September 3, 1929, adjusted high prices**
10	November 13, 1929

* Except RCA stock, which is listed at the high for the year.

** Reflects some splits and options not included in the game.

Name ______________________________

Date ______________________________

WORKSHEET

NAME OF STOCK SOLD	PRICE	NO. OF SHARES	TOTAL COST	MARGIN LOAN	CASH
NAME OF STOCK BOUGHT	PRICE	NO. OF SHARES	TOTAL COST	PAID FOR	CASH HELD
				CASH: BORROWED: TOTAL:	
HELD AT END OF DAY: NAME OF STOCK	PRICE	NO. OF SHARES	BORROWED ON MARGIN		CASH HELD

Name ______________________________

Date ______________________________

NAME OF STOCK SOLD	PRICE	NO. OF SHARES	TOTAL COST	MARGIN LOAN	CASH
NAME OF STOCK BOUGHT	PRICE	NO. OF SHARES	TOTAL COST	PAID FOR	CASH HELD
				CASH: BORROWED: TOTAL:	
HELD AT END OF DAY: NAME OF STOCK	PRICE	NO. OF SHARES	BORROWED ON MARGIN		CASH HELD

Name ______________________________

Date ______________________________

SAMPLE WORKSHEET

NAME OF STOCK SOLD	PRICE	NO. OF SHARES	TOTAL COST	MARGIN LOAN	CASH
NAME OF STOCK BOUGHT	PRICE	NO. OF SHARES	TOTAL COST	PAID FOR	CASH HELD
GM	10	100	$1000.00	CASH: $1000.00 BORROWED: TOTAL:	0
HELD AT END OF DAY: NAME OF STOCK	PRICE	NO. OF SHARES	BORROWED ON MARGIN		CASH HELD
GM	10	100	0		0

Name ______________________________

Date ______________________________

NAME OF STOCK SOLD	PRICE	NO. OF SHARES	TOTAL COST	MARGIN LOAN	CASH
NAME OF STOCK BOUGHT	PRICE	NO. OF SHARES	TOTAL COST	PAID FOR	CASH HELD
GM	10	1000	$10,000	CASH: $1000.00 BORROWED: $9000.00 TOTAL: $10,000.00	0
HELD AT END OF DAY: NAME OF STOCK	PRICE	NO. OF SHARES	BORROWED ON MARGIN		CASH HELD
GM	10	1000	$9000.00		0

Chapter XII

Berlin, 1948

When the Soviet Union broke up, the Cold War era ended. While it's a relief no longer to be facing the constant tension and threat of nuclear war, it's also a problem. How can students who never experienced it understand the Cold War? My students have asked, "Why is there always this negativism about communism?" The conflicts in Korea and Vietnam are big enough to register with the students, but what about the less violent pressures that were unrelenting during the Cold War?

The Berlin, 1948, game is an attempt to put students inside a nonmilitary Cold War crisis. This game is good for classes of any size. It is part briefing, which is a short lecture, then small-group discussion, followed by class discussion. This makes it a pretty sedate game.

Preparation

The teacher should make a transparency from the master at the end of this chapter and then review the briefing and prepare to deliver it. I use groups of two students in the discussion phase. You may use larger groups. It's your classroom, and you know what works best. This game should be played before the class has read about the blockade so they approach the problem with open minds.

A Sample Game

The briefing.

Today we are going to play a simple game involving the Cold War and Berlin. For this game each of you will be President of the United States. You will be asked to make a policy decision about a crisis, a real crisis that we faced in 1948. You are playing President Truman. After a factual briefing, I will list your policy options. Listen carefully.

Recently the Soviet Union has been very difficult. When World War II ended, we demobilized our armed forces and disarmed. The Soviets did not. They kept a large army in Eastern Europe and used it to force the countries of Eastern Europe to become Communist. We see this as a threat to Western Europe and to our security. Germany is a special case. Since they were the enemy during World War II, they were divided into zones of occupation. The four Allied powers—United States, United Kingdom, France, and the Soviet Union—each occupied a portion of Germany. This map shows these zones. (At this point show the map of Germany.) Notice that Berlin is located in the Soviet zone, approximately a hundred miles away from the American, British, and French zones. Berlin itself is divided into four zones and is run by a joint council of the Allies. The division of Berlin is shown in the Berlin Occupation map. Until recently, residents of the American, British, and French zones in Berlin have been supplied by road, railroad, and canal from the American, British, and French zones of Germany. The Soviet Union has just closed those surface supply routes. The Soviets are also preventing all movement of supplies into Berlin. They used to provide electricity, but that is now cut off. At the moment, the only link open to Berlin is by air. There are airports in our zone of Berlin, and we have air corridors 20 miles wide through which we can fly.

We are convinced that this blockade is a Soviet attempt to take over all of Berlin. By doing so, they would add over two million people to their empire. We face a dilemma: The people of Berlin need roughly 4,500 tons of supplies each day. This includes food, clothing, medical supplies, and coal for power and heat. Current supplies in Berlin will last roughly four weeks. What should we do? Here are some of the policy options you may want to consider. (At this point show the list of policy options.)

1. *Give Berlin to the Soviet Union.* The city is surrounded by Soviet territory, and their forces in the area are already much stronger than ours. Our British allies are opposed to this option, and U.S. representatives in Berlin and Germany have already said we will not abandon Berlin. A change now would destroy our credibility.

2. *Threaten war with the Soviet Union.* Our analysis of forces on the ground indicates that we would lose a ground war. We do have the atomic bomb and bombers capable of dropping it on Soviet cities. We do not have B-29 bombers based in Europe that can deliver atomic bombs, but we could deploy them shortly.

3. *Send in an armored column to escort supplies to Berlin.* General Clay has a plan to send in a convoy. He would assemble a force of about 5,000 men with tanks and even bridge-building equipment and push up the autobahn (the German main highway). General Clay expects the Soviets to put up some barricade, but he does not expect them to fight. He could get about 500 tons of supplies through. He sees this as the most practical alternative.

4. *Airlift.* Airports and air corridors are available. We do not know if the Soviets will use their fighter planes to stop our transports. The risk of confrontation certainly seems less than with the armored column. However, there is no precedent for an air supply effort of this size. It would require us to strip our armed forces of all cargo planes. This would leave us exposed if we are threatened in other areas. We can draw on our experience flying supplies "over the Himalayas" to China during the war.

"What is your decision, Mr. President?" At this point, I direct students to pair up with their partners and discuss the pros and cons of each option, with each then deciding what to do. I allow them about five minutes. I routinely pair up students sitting next to each other. They simply turn toward each other and start talking. The room buzzes with voices. After five minutes I say, "Now each of you, get out paper and write down your decision. Give the reasons you chose that option." More buzzing as some students borrow paper or pencils is followed by quiet as they write their choices and their reasons.

Depending on the situation, I may or may not collect these papers. If class is nearly over, I usually collect them so I can read and grade them before debriefing. Obviously, if I have a backlog of correcting, I let students keep their papers.

Debriefing

"Now, let's see what you decided. Bill, what was your decision?"

"I decided to use the airlift."

"Why?"

"It seemed to me that it was less risky than the armored column, and we didn't want either a war or to give up Berlin."

"Thank you, Bill. Sue, how did you decide?"

"I picked the airlift, too. It seemed to me the armored column had a good chance of starting a war. The other options were really not choices we could accept. That left airlift, even if there was a chance it wouldn't get enough supplies in. None of the options looked very good to me."

"Did anybody pick another option?" I ask. "If so, raise your hand."

So far my students have nearly all chosen the airlift. Perhaps this reflects a bias in the presentation. Also, today's students have grown up with air freight and airlifts. They don't see the airlift option as the challenge it was in 1948. If there is another choice, it is the armored column. I can't recall any student supporting the other two options. They are pretty clearly unacceptable. The armored column may have been less risky than it looked in 1948. Soviet insiders have said that Stalin did not want war over Berlin. Presumably Soviet forces would not have used force against the column. But also it would have taken nine such columns a day to deliver the supplies Berlin needed.

President Truman chose the airlift. Soon, the people of Berlin were receiving all their supplies by air. The general who had managed flying supplies over the Hump to China was put in charge, and he organized a very efficient system. There was still danger, however. Soviet fighter aircraft could attack at any time. The cargo planes were unarmed. Now, of course, we know that no attack took place, but in 1948 the threat of attack was part of the constant tension during the Cold War.

Bibliography

Truman, Harry S. *Memoirs: Years of Trial and Hope*. Vol. II. Garden City, NY: Doubleday, 1956. (The game follows President Truman's account of the Berlin crisis.)

Tusa, Anna and John. *The Berlin Airlift*. New York: Atheneum, 1988.

Name ____________________

Date ____________________

BERLIN OCCUPATION

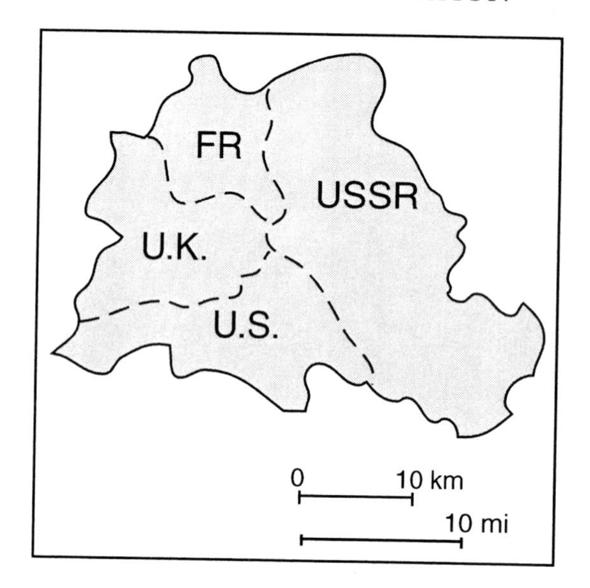

OPTIONS

1. Give Berlin to USSR
2. Threaten war
3. Armored column
4. Airlift

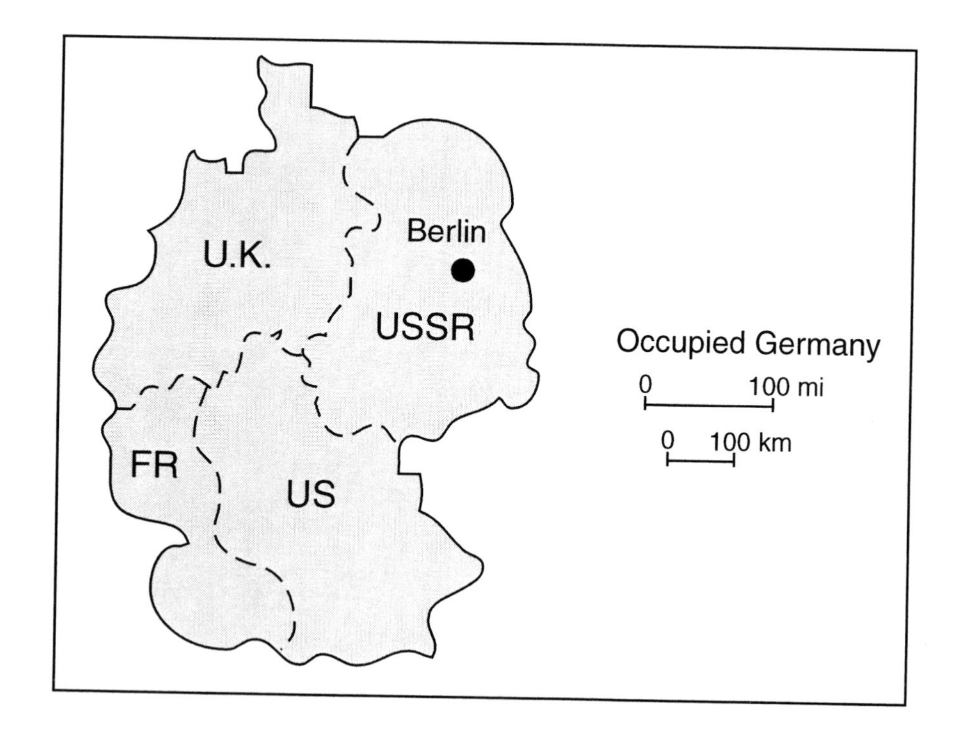

CHAPTER XIII

GULF OF TONKIN

When I teach about the Vietnam War, one of my students usually asks the very basic question, Why did we ever get involved? The answer is a mixture of motives, misunderstandings, and mistakes. The stated guiding principle of American foreign policy at the time was simple: resist communist expansion wherever and whenever it occurred. Some U.S. leaders wanted to roll back communism. A small "brush fire war" in a distant Asian country was just one of many perceived challenges from the "worldwide Communist conspiracy." The United States attempted to deal with it and became entangled step by step in Vietnam. Later, most Americans decided that our perception of the situation in Vietnam did not match reality.

This exercise puts students at a critical turning point and asks them to make a decision. It is designed to give a feel for the attitudes of the time. It is best to play the game before assigning readings on Vietnam so your students will approach the decision with open minds. I have changed the names of North and South Vietnam to try to avoid giving the game away, but all other names are historic.

PREPARATION

You need to make enough copies of each handout to have one full set for each group of four students. You can either include the map or make a transparency of it and project it on an overhead. You don't need to prepare student in advance. The exercise needs about a full class period.

Do not give the game procedure sheet to your students: It is your lesson plan. Note that you have some choices to make in step 3. Option A is the simplest. You simply hand out the information to each student and let them read it; then the "senator" and staff meet to discuss and arrive at a decision. Option B is similar, except that before the senator and staff meet, students in a, b, and c roles meet in groups to discuss their information. In other words, all students doing context briefing meet, as do the McNamara-Rusk briefing group and the resolution and interpretation group. The senators play golf. Then each senator meets with staff to discuss a decision. Option C allows senators to debate their decision before voting. You can simply call the class to order and ask whether any senator has something to say before the vote. You then recognize senators in turn. You may want to set a time limit for the debate. Once debate is over, you call for a vote. When the vote is recorded and the senator has written the reasons for the debate, you can start debriefing. You may adjust the game schedule and procedure to fit your classroom.

ORIENTING THE CLASS

To start the game, say something like "Today we will be playing a political decision game set in the Cold War. I will put you in groups of four, and one member of each group will be a senator who must make a decision. First we need to assign roles. Please count off a, b, c, d." Each student is assigned a letter. Any extra players after the class has broken into fours become senators and borrow staff. In a class of 30, for example, there would be seven groups of four and two extras who become senators. They sit in with one of the groups, vote, and make decisions, and so forth. If one of your a, b, c players is weak, you could use an extra to help out. After every student has a role, you distribute the appropriate handouts, a ballot to each senator and briefing material to the others. Explain that there is a crisis and that each student has different information or responsibilities; then allow time for them to read their handouts. At this point you may also want to show them an outline of how you want them to do step 3. You are now off and running.

A Sample Game

I ran this game in a small-group session at the Iowa Council for the Social Studies Conference. The teachers immediately recognized the situation, but they were excellent role players. When the game was under way, I overheard dialogue like, "How was your golf game, Senator? Did your secretary accompany you, as she does most afternoons?" When I polled the senators, a bare majority voted for the resolution. Hindsight is a powerful influence. Those who voted against it argued that the resolution granted too much authority. By contrast, high school students who have played the game did not recognize the situation and voted for the resolution by huge margins.

Debriefing

You can play the game, record the results, and save the debriefing until the Vietnam War comes up for study. Debriefing could also be a transition to begin studying the Vietnam War. Questions for debriefing are: Why would anybody support this resolution? What were some possible problems with it? Why was fighting communism so important at that time? What was the role of election-year politics in this decision? At some point, you need to explain to students that this is an actual resolution that passed the Senate by a vote of 88 to 2 and the House by 416 to 0. Only two names have been changed: Anam represents South Vietnam, and Tonkin is North Vietnam. The dates, time, and reports in the game are as they were at the time.

Later, this resolution became controversial, as public opinion turned against the war. In 1968, the Senate Foreign Relations Committee concluded that there was no attack on August 4. Robert McNamara, Secretary of Defense in 1964, later disagreed. The destroyers in question were on missions to gather electronic intelligence, recording radio transmissions and picking up on radar beams from North Vietnam.

Bibliography

McNamara, Robert. *In Retrospect: The Tragedy and Lessons of Vietnam.* New York: Time Books, 1995. (This is the source for chronology and briefings.)

Commager, Henry S. *Documents of American History,* Vol II. Englewood Cliffs, NJ: Prentice Hall, 1988. (For the full text of the resolution, see page 689 in this reference.)

GAME PROCEDURE
(DO NOT HAND OUT)

1. Make a short statement to set the scene.
2. Assign roles, one player to each.
   (a) Context briefing
   (b) McNamara-Rusk briefings
   (c) Resolution and interpretation
   (d) Senator
3. Select one option from the following:
   (a) Form groups of four and have them exchange information, discuss it, and reach a decision.
   (b) Run a jigsaw exercise. Have each expert adviser (a, b, c) meet with peers from other groups and then reassemble in decision groups. Problem group "d" does nothing (plays golf) until meeting with a, b, c.
   (c) Have groups of four meet and recommend a decision, then allow a "debate" among the senators before the final vote.
4. Collect papers.
5. Select one option from the following:
   (a) Debrief the same day or in the next class.
   (b) Postpone debriefing until Vietnam comes up in class.

Name ______________________________

Date ______________________________

GULF OF TONKIN

CONTEXT

It is 1964 and the United States is deeply involved in the Cold War. This struggle against the Soviet Union is the biggest concern of American leaders and voters. President Lyndon Johnson is in the early stages of a reelection campaign. His opponent, Republican Senator Barry Goldwater, has taken the position that the Democrats are too soft on communism and that we should use military force to fight Communists.

One of the issues in the election is U.S. policy toward Anam. Anam is a small Asian country being attacked by guerrillas sponsored and supported by Tonkin, a neighboring Communist state. The guerrilla attack is supported by the two major Communist powers, the Soviet Union and China. The United States believes that if Anam falls to communism, then other Asian countries will fall in turn, like dominoes. Therefore, the United States is providing aid to the government of Anam. There are now 16,000 American soldiers serving as "advisers" in Anam. President Johnson wants to contain and limit the conflict in Anam, to avoid the possibility of war with the Soviet Union or China. Senator Goldwater favors expanded use of military force.

Recent news reports say American ships were attacked by Tonkin's ships. As a senator you will be asked to cast a vote that will shape U.S. policy.

BRIEFINGS

First Briefing: August 3, 1964

By: Secretary of Defense McNamara and Secretary of State Dean Rusk

To: Senate Foreign Relations Committee

Senate Armed Services Committee

On August 2, 1964, at 3:40 P.M. Anam time, the U.S. destroyer *Maddox* was patrolling in international waters 25 miles off the coast of Tonkin. Torpedo boats from Tonkin approached and fired torpedoes and automatic weapons. There was no damage, but shell fragments did land on the *Maddox*. The President decided not to retaliate. He is sending a note of protest to Tonkin and is adding another destroyer, *C. Turner Joy,* to the patrol.

Second Briefing: August 6, 1964

By: Secretary of Defense McNamara and Secretary of State Dean Rusk

To: Senate Foreign Relations Committee

Senate Armed Services Committee

On August 4, 1964, at 7:40 P.M. Anam time, *Maddox* radioed that an attack was imminent. They had picked up three vessels on radar. Fighter aircraft from the carrier Ticonderoga were launched. This was a night of thunderstorms, low clouds, and no moon. In the following hours, the *Maddox* and the *C. Turner Joy* reported over 20 torpedo attacks, sighting torpedo wakes, enemy cockpit lights, searchlight illumination, and automatic weapons fire, as well as radar and sonar contacts.

There were some initial doubts about the accuracy of these reports, but the patrol commander has since sent a follow-up message that the original ambush was bona fide. The President has decided to make a limited response. U.S. carriers have launched a bombing attack against patrol boat bases and a supporting oil complex.

Name ______________________

Date ______________________

RESOLUTION AND INTERPRETATION

This resolution has been introduced in the Senate:

Whereas naval units (of Tonkin) . . . in violation of international law, have deliberately and repeatedly attacked United States naval vessels lawfully present in international waters . . . and whereas these attacks are part of a deliberate and systematic attack against its neighbors . . . the United States is, therefore, prepared, as the President determines, to take all necessary steps, including the use of armed force, to assist any member or protocol state of the Southeast Asia Collective Defense Treaty requesting assistance in defense of its freedom.

Statement by Secretary of State Dean Rusk:

". . . we cannot tell what steps may in the future be required . . . as the Southeast Asia situation develops, and if it develops in ways we cannot now anticipate, of course there will be close and continuous consultation between the President and Congress."

On the Senate floor, Senator John Sherman Cooper (R-Kansas) asked Senator William Fulbright (D-Arkansas) a series of questions.

Cooper: Are we now giving the President advance authority to take whatever action he may deem necessary respecting Anam and its defenses, or with respect to the defense of any other country included in the (SEATO) treaty?

Fulbright: I think that is correct.

Cooper: Then looking ahead, if the President decided that it was necessary to use such force or could lead us into war, we will give that authority by this resolution?

Fulbright: That is the way I would interpret it.

THE SENATOR

Voting: When the resolution comes to a vote, you will vote yes ______ or no ______.

In fifty words or less, justify your vote.

Chapter XIV

Presidential Election

Presidential politics is the great American game that millions of people play, not merely as voters, although that is a rare privilege in the world, but as campaign workers, delegates, and for a select few, as candidates. The game lasts roughly two years. The candidates start setting up their organizations in the key early primary or early caucus states right after the off-year election. In January of the election year, the Iowa caucuses start it off. From Iowa it's off to New Hampshire and then through a grueling series of contests until the parties select their nominees. There may be a breathing spell until Labor Day; then the campaigning gets serious. At last, on election day, the voters have the final say in the matter, and things calm down for a while, except that a few fanatics are already looking toward the next election. All this action is reported and analyzed on television, radio, in print, and on the Internet.

Nominating and electing a president offers social studies teachers a chance to take advantage of the media hype to get students involved. Elections are a great opportunity to focus on education for citizenship.

As teachers of United States history know, procedures for nominating and electing a president have gone through many changes. Political parties did not hold nominating conventions until the middle of the nineteenth century, but ever since national conventions have been the primary public events in every campaign. It was conventions that were the sites for the "Cross of Gold" speech and the first "smoke-filled room."

In the last half of the twentieth century, state primaries and state caucuses or conventions have become the decisive contests, although they are far from being the prime media events. National conventions only make a nomination that has already been decided official. Every candidate since the 1940's has been nominated on the first ballot. Any drama in conventions since then has been about platform issues or choice of vice-presidential candidates. And, some conventions have been very dramatic. In 1948 the Democrats actually split over a civil rights platform plank. In 1964 the Republicans argued angrily over "extremism." Lyndon Johnson's acceptance of the vice-presidential nomination was a shock, as was George Bush's choice of Dan Quayle as a running mate.

But more and more, the contest and much of the drama take place on a state-by-state basis, before the convention. The 1996 presidential conventions were slick, stage-managed infomercials. All decisions had been made and announced in advance. Both parties concentrated on projecting an image that they hoped would help win votes. Paradoxically, television viewers stayed away from these events made for television. It's not much fun watching hours of political infomercial. If future conventions are like those in 1996, there is not much point trying to use conventions as the basis for simulations. There were few real debates, no real excitement, and no decisions, except, of course, for viewers' decisions to watch reruns on another channel.

So, social studies teachers looking for a decisive moment in the campaign needs to look at the contests before the convention. The caucuses and primaries have suspense, surprises, drama, and humor. They are also easily adaptable to classroom simulations. The media coverage is adequate for students to research issues relatively easily, and the events are governed by a few simple rules.

Iowa Caucuses

In recent presidential campaigns, the first major media event has usually been the Iowa Caucuses. The 1996 Iowa Caucuses were held in February. In the next election year they may be held in January, and perhaps some other state may challenge Iowa for the status of first contest, but it still looks probable that in 2000 Iowa will again be first. The contest has already started.

In fact, the contest for nominations in 2000 started at the 1996 conventions. There was not much for delegates to do besides provide a living background for the conventions. They went to meetings a lot. A number of out-of-state politicians dropped by the Iowa delegation to say something like, "just keeping in touch with my Iowa friends." A friend who is a convention delegate can be very helpful to candidates building a campaign organization for the caucuses. That was just the preliminary state. Serious candidates visit Iowa and kiss a pig. Pork is big business in Iowa. Candidates who want to win in our caucuses need to visit a farm and be photographed holding a pig in their arms. (I exaggerated about the kissing.) This and other indignities are a routine part of the campaign to make a decent showing in the Iowa caucuses.

What are caucuses, you ask? Good question. They are an event for political junkies and an opportunity for social studies teachers. I qualify on both counts. I teach in Iowa and have been attending caucuses since 1966. I believe I have missed only one during that period. I attend Democratic caucuses. The Democrats have more complex rules than the Republicans, but the general organization is the same. Caucuses are held each election year in every precinct in Iowa. The precinct is the smallest political unit in Iowa. It is the area from which people come to one polling place to vote. There are about two thousand precincts in the state.

Caucuses are held in schools, church basements, and living rooms. An off-year caucus can be as small as three or four people. In presidential election years, caucuses can swell to over a hundred. Iowa weather can be very hostile: Subzero cold or a blizzard discourage attendance. Caucuses are relatively informal, but some rules must be strictly followed. Official starting time is 7:00 P.M., but delegates cannot be chosen until 7:30, which allows for latecomers. The Democrats have a rule about selecting delegates by preference groups. In practice, it is complex and allows candidates with less than a majority to pick up a few delegates. The delegates in this case are being chosen to attend the county convention, where a delegation will be chosen for state and district conventions. At these levels, delegates are chosen for the national convention. Note that the link between a caucus decision and national delegate selection is pretty weak. It is absolutely routine for delegates chosen at a caucus to find that their candidates have withdrawn by the time of the county convention.

The caucus also passes a list of resolutions that are sent on to the county convention for consideration as part of the county platform. This process starts during the first 30 minutes of the caucus, often stops by agreement when it's time to choose delegates, and then resumes.

The caucus also names two precinct committee members and signs nominating petitions. This is actually its main business, but the presidential campaign gets all the attention. And the next day the media declares a winner and heads off to New Hampshire.

Orienting the Class

In my caucus simulation I organized everything a week or so beforehand. I prepared a list of candidates—in 1996 the Republicans had nine candidates—and set up a a team for each: one student to prepare a two-minute nominating speech, another to write a platform statement that the candidate would be likely to support. For example, the Steve Forbes team supported the flat tax. It was their candidate's key issue. They needed only one idea. A third team member was to prepare a poster for the candidate. I specified that posters be no larger than 8" × 12" so I could fit them all on my bulletin boards. I reserved a class period for the simulation on the day of the actual Republican caucus.

A Sample Game

On the day of the simulated caucus, I preside and use a simplified agenda. I just call the meeting to order and start the nomination of candidates. Each designated speaker is recognized and speaks for his or her assigned candidate. Some are pretty smooth; others stumble. Some teams have no speechmakers, but I insist that some member of the team at least say, "I nominate ________." Once the names of all candidates are before the caucus, I simply call for a vote. Students are free to vote as they feel, not for the team they were assigned. And, since this is a Republican caucus, the candidate who receives the highest number of votes wins all the delegates. In a typical precinct this would be six. After that, the class moves on to adopting platform resolutions. We do this until the class period ends.

I encourage my students to attend caucuses. Both parties make a point of welcoming young people. The Republicans have a "junior delegate" system for Republicans too young to vote. When I taught government, some of my students were eligible voters, or would be by election day. Some of them were elected to precinct committees, and others became delegates. In Iowa both parties are very open about people attending caucuses. Any eligible voter who is willing to sign a statement that he or she is planning to support the party's candidate can participate in the precinct caucus. I have known active Republicans who attended a Democratic caucus when the Democrats were having the contested nomination.

Debriefing

The next day is a regular current events day. It is also a chance to compare the simulated caucus with the real thing. Did any of you attend the real caucus? What was it like? Who did your caucus support? Were there any arguments about platform? How close to reality was our simulation? What is the media saying about yesterday's caucuses? Who are they calling the winner? Were there any platform issues that made news over the whole state? Who has been hurt by caucus results? Has anybody dropped out of the race? Why are organized groups so important in the caucus process? What is the next contest for these candidates? What do the polls show about candidate strength there? I schedule another current events day on the day after the New Hampshire primary.

Alternatives

All this is very nice if you teach in Iowa. What about other states? My impression is that nationwide media coverage of the Iowa caucuses is intense enough that the Iowa caucuses can be the basis for a simulation followed by a current events day analyzing the real thing. From place to place how much coverage your students are exposed to may vary.

Perhaps your state is a primary state or is close enough to get media spillover from a primary state. In that case, you can do a primary simulation. These are even easier than caucuses. For the most basic exercise, all you have to do is make up a ballot that lists possible candidates and have your students vote. The ballot can be a paper ballot which you pass out in class, or it could be a computer polling station set up to register votes. Election officials might even make an actual voting booth available. A call to the city hall or county courthouse will put you in contact with the proper officials. Their titles vary from state to state. My high school routinely gets sample ballots and dummy voting machines from our election officials.

In a primary election, the candidates are listed on separate ballots by party, and voters get only a contested nomination. Obviously, that's the one to simulate. In 2000 there may be open races in both parties. How do you handle that? I suppose we wait and see. If one contest is livelier, go for that party.

It's not possible to set up a game in advance that will apply in every state in every election year. States keep changing the dates of caucuses and primaries; rules for the process change a bit each time, and, of course, each contest is different. But if you use the ideas above, you can plan a classroom simulation that will work for your students. I have provided a sample primary ballot and an outline for an Iowa Caucus simulation. There is also a list of delegate strengths for the states in 1996 for both parties.

In the nominating process, it's the delegates who really count. When a candidate has a majority of the delegates, the contest is over. That's why it's smart to do simulations of the early stages of the process.

The conventions are held during the summer, so when school resumes in the fall, the focus is on the general election. Simulated general elections are easy. You can have students vote in your classes, or perhaps the school will opt to hold a school-wide election. During election years I set up two bulletin boards, one for each of the two parties. I then invite my students to bring in campaign posters for their party. We follow polls and campaign issues on regular current events days.

In the general election, there is a good chance that your students can be directly involved. Both parties are usually looking for workers. The jobs are not glamorous. They need people to put up yard signs, paste on address labels, make calls to voters, and do other routine but vital tasks. My high school allows extra credit in social studies class to students who spend ten hours working for a political party. You may want to consider some such optional assignment. Education for citizenship is a major purpose of social studies. Giving students a chance to simulate the election process can be a powerful tool to prepare them for participation in government.

Schedule: Iowa Caucus

1. Two weeks before caucus: Brief students on what a caucus is. Set up teams for candidates.

   (a) Candidate teams of three students: a person to nominate, a platform person, a poster person

   or (b) Candidate teams of a person to nominate, a slate of delegates (six), and a number of platform writers.

2. Day of caucus: Caucus "Anytown, Iowa" with six delegates going to county convention.

   Agenda

   (a) Nomination speeches.

   (b) Voting for candidates.

   (c) Voting on platform issues.

3. Caucus night 7:00 P.M.

   Attend your caucus. Watch the TV reporting of the caucus.

4. Day after caucus: Current events day on caucus results and debrief comparing "our caucus."

Bibliography

Books

Goldstein, Michael L. *Guide to the 1996 Presidential Election.* Washington: Congressional Quarterly, Inc., 1995

Ragsdale, Lyn. *Vital Statistics on the Presidency.* Washington: Congressional Quarterly, Inc., 1996.

Web Sites

Gallup Organization. http://www.gallup.com/

Iowa Cyber Caucus. http://www.drake.edu/public/news.html

University of New Hampshire. http://www.unh.edu/ (There was no material on the New Hampshire primary when I checked it in July 1997, but in 1995 it had a site on the 1996 primary.)

1996 Calendar of Delegate Selection

Date	State	Method	Democratic	Republican
			Delegates chosen	
Jan. 25–31	Hawaii	Caucus		14
Jan. 26–29	Alaska	Caucus		19
Feb. 6	Louisiana	Caucus		28
Feb. 12	Iowa	Caucus	48	25
Feb. 20	New Hampshire	Primary	20	16
Feb. 24	Delaware	Primary		12
Feb. 27	Arizona	Primary		29
	North Dakota	Primary		18
	South Dakota	Primary		18
March*	Nevada	Caucus		14
	Wyoming	Caucus		20
March 2	South Carolina	Primary		37
March 3	Puerto Rico	Primary		14
	Virgin Islands	Caucus		4
March 5	American Samoa	Caucus		4
	Colorado	Primary		27
	Connecticut	Primary		27
	Georgia	Primary		42
	Idaho	Caucus	18	
	Maine	Primary	23	15
	Maryland	Primary	68	32
	Massachusetts	Primary	93	
	Minnesota	Caucus	128	33
	Rhode Island	Primary	22	16
	South Carolina	Caucus	43	
	Vermont	Primary		12
	Washington	Caucus	74	36
March 7	New York	Primary	244	102
	Missouri	Caucus		36
March 9	Missouri	Caucus	76	

1996 Calendar of Delegate Selection

Date	State	Method	Democratic	Republican
			Delegates chosen	
March 9	Arizona	Caucus	44	
	South Dakota	Caucus	15	
	Democrats Abroad	Caucus	7	
March 10	Nevada	Caucus	18	
	Puerto Rico	Primary	51	
March 12	Florida	Primary	152	98
	Hawaii	Caucus	20	
	Louisiana	Primary	59	28
	Mississippi	Primary	38	32
	Oklahoma	Primary	44	38
	Oregon	Primary	51	23
	Tennessee	Primary	68	37
	Texas	Primary Caucus?*	194*	123
March 15	Delaware	Caucus	14	
March 16	Michigan	Caucus	128	
March 17	Puerto Rico	Primary		14
March 19	Illinois	Primary	164	69
	Ohio	Primary	147	67
	Wisconsin	Primary	79	36
March 23	Wyoming	Caucus	13	
March 25	Utah	Caucus	24	28
March 26	California	Primary	363	163
March 29	North Dakota	Caucus	14	
March 30	Virgin Islands	Caucus	3	
April 2	Kansas	Primary		31
April 13	Alaska	Caucus	13	
April 13/15	Virginia	Caucus	79	
April 23	Pennsylvania	Primary		73
May 4	Kansas	Caucus	36	
	Guam	Caucus/Convention	3	

1996 Calendar of Delegate Selection

Date	State	Method	Democratic	Republican
			Delegates chosen	
May 7	District of Columbia	Primary	17	14
	Indiana	Primary	74	52
	North Carolina	Primary	84	58
May 14	Nebraska	Primary	25	
	West Virginia	Primary	30	
May 21	Arkansas	Primary	36	20
May 28	Kentucky	Primary	51	26
June 4	Alabama	Primary	54	4
	Montana	Primary	76	48
	New Jersey	Primary	104	48
	New Mexico	Primary	25	18

*no date

Notes:

1. For reasons too complicated to go into here, the Democratic delegate numbers are not the total number of delegates in the state delegation. They are those chosen by the process.

2. In some states the Democrats and Republicans hold their contests on different dates. Those states are listed under both dates.

3. Some of these 1996 dates and delegate strengths will change for 2000 and beyond.

Sources: The Democratic National Committee, 430 South Capitol Street, S.E., Washington, DC., and *Guide to the 1996 Presidential Election.* Michael L. Goldstein, Congressional Quarterly, Inc., Washington, D.C., pp. 30–31.

Name ______________________

Date ______________________

ELECTORAL VOTES—2000

State	Votes
Alabama	9
Alaska	3
Arizona	8
Arkansas	6
California	54
Colorado	8
Connecticut	8
Delaware	3
Florida	25
Georgia	13
Hawaii	4
Idaho	4
Illinois	22
Indiana	12
Iowa	7
Kansas	6
Kentucky	8
Louisiana	9
Maine	4
Maryland	10
Massachusetts	12
Michigan	18
Minnesota	10
Mississippi	7
Missouri	11
Montana	3
Nebraska	5
Nevada	4
New Hampshire	4
New Jersey	15
New Mexico	5
New York	33
North Carolina	14
North Dakota	3
Ohio	21
Oklahoma	8
Oregon	7
Pennsylvania	23
Rhode Island	4
South Carolina	8
South Dakota	3
Tennessee	11
Texas	32
Utah	5
Vermont	3
Virginia	13
Washington	11
West Virginia	5
Wisconsin	11
Wyoming	3
District of Columbia	3
Total	**538**

CHAPTER XV

REPORTS

The oral report is an excellent and trusted standby in social studies teaching. It gives students a chance to practice research and communication skills, and it offers a change of pace.

Unfortunately, oral reports can also be dull, and more confusing than helpful. Few classroom experiences are worse than sitting through a series of bad reports. One after another, students move to the front of the room and read in a monotone from a copy of an encyclopedia article. They stumble over words and rarely look up at their audiences. When they've finished, they can't even respond to a simple question like, What was the main idea of this report?

I would like to tell you how I have inspired all my students to give really great reports, but it wouldn't be true. Because I teach a normal range of high school students, some reports are very good, most are average, and some are very bad. What I can share is my system for assigning oral reports that gives every student a fair chance to do a good job. I can also share some types of reports that are interesting even if they fail.

THE SYSTEM

Perhaps the best way to describe my system of assigning reports would be to let you listen in on a telephone conversation I once had with a rather upset mother.

"Mr. Scott, I think it wasn't fair of you to give Jose such a bad grade on his report. That was a very hard topic for him."

"Well, he didn't have to pick that topic. There were about forty topics on the project list."

"But he was among the last to choose. By that time only about ten topics were left."

"True. But next time he'll be earlier. I always rotate so that no student is first or last every time. Did he consider suggesting a topic not listed? That's always an option. Several students did that."

"I didn't know about that. It's too bad he didn't have a chance to see the list before he signed up. Maybe he would have realized the possibility."

"Odd you should mention that. At the end of the unit test, most students had some time, so I let them see the list. I have several copies, one per student. We did the sign up the following day. Jose wrote a very short essay. He had plenty of time to look at the list."

"Oh. Well, I still think the topic was too hard for him. He just couldn't find a thing on it."

"Did you read my guarantee?"

"What guarantee, Mr. Scott?"

"The guarantee at the bottom of the assignment sheet that says, 'If you are unable to find material for your report, contact me at least 24 hours before the report is due. If I cannot find material, you can drop that project and select a new one.'"

"I never noticed that. I'm going to talk with Jose about his reports. I wish you would give him better grades in the future. He's such a good boy, a real leader in the family."

"I hope he'll earn better grades in the future, Mrs. Diaz."

"Goodbye, Mr. Scott."

"Goodbye."

Let's look at the system in a more formal way. The project list is long and varied, about forty projects, including some of the variations I will discuss later. Students always have the option of suggesting a topic. (In practice, this usually means that the student talks with me and we jointly work out a topic for a project. Students usually have the opportunity to look at the project list the day before signing up. Students sign up for projects in a different order each time, so no one is always first or last. Projects are due when they fit best with the reading and discussion topic for the day. I encourage students to start their research early. If they are unable to find material, I am available to help them. They are expected to prepare their reports during study hall or other free time. I encourage students to deliver reports from outline notes, just as they would deliver a speech to persuade or inform. When the report is over, I encourage the class to ask questions; then I ask a few questions. Students get a grade and a short set of written comments after the report.

Flavor of History

The assignment I call Flavor of History lets students chew on and digest some American history. Usually the assignment sheet simply says, Prepare a recipe from this period and bring it to class. If a student signs up for this project, I usually add a few oral instructions: "Try to be as authentic as possible. Get a genuine recipe from a reliable source. Avoid substituting ingredients. Be sure to tell us why the recipe was prepared the way it was. Remember, the point of this report is to use food to understand how people lived. You don't have to feed everyone huge servings; just a taste will do."

Research for this report takes us into cookbooks. There are more of them than I want to count, including some primarily concerned with old recipes and some that just have a few old-fashioned recipes. In my area many of the towns were founded in the 1880's, so centennial cookbooks with some old-time recipes are relatively easy to find. Outdoor magazines often have recipes for foods that keep well without refrigeration, like old-fashioned chili, bannock, or jerked beef. A short list of possible sources appears in the bibliography.

Once a student has located an authentic recipe, the next problem is ingredients. Some items are simply no longer available, and substitutions are the only choice. The important thing is that students notice the substitutions and tell the class that they have made them.

Modern-day heat sources are usually not authentic, so students should make an effort to approximate original conditions. Ramrod bread from the Civil War, for example, should be a strip of dough wrapped around a stick and held over an open fire. I don't insist on an actual ramrod, but I do require an actual open fire. Authentic Plains Indian or pioneer cooking used a fire made of buffalo chips, not wood. Buffalo chips are dried manure. I understand they burn cleanly with no distinctive smell. But, since buffalo are in short supply, we haven't had to prove the point. Most of our recipes are immigrant cooking or nineteenth-century farm cooking, so usually a stove is acceptable. Modern gas or electric ranges are allowed to substitute for wood cookstoves.

The critical point comes after we have finished our sourdough biscuits or Indian fried bread on a napkin. Now the student cook has to explain how this recipe reveals something about American history. One case stands out in memory. A student brought in something he called "onion pie," made from an old family recipe going back to his Italian grandfather—only since his grandfather's time, the recipe had undergone some modification to become indistinguishable from modern pizza. He went on to explain that onion pie (pizza) is a sort of quiche. Not all efforts are so successful. I remember the sourdough chocolate cake. The student chose chocolate to override any sourdough flavor! Another student's family didn't like the stew he had prepared from an authentic recipe, so, like a born chef, he doctored it up to taste better! Mostly, we eat a lot of bread. It's an easy food to transport and serve and a staple in every cuisine: a good, practical choice. Most student reports simply explain what conditions led to the choice of ingredients in their recipes. I notice that my classes always look forward to a Flavor of History report. I just hope it gives them food for thought.

Artifacts

American history is everywhere. You walk on it, live in it, and sometimes trip over it. It survives in our streets, sidewalks, houses, and in items that we use every day. If you collect enough of almost any artifact, you can charge admission—arrowheads, dishes, furniture, cars, or even old spacecraft. You can find American history in Colonial Williamsburg, in Old Fort Bliss, or in Grandmother's old sugar bowl. Artifacts draw crowds because they connect daily life, past and present. They are inherently interesting wherever they turn up, including in your classroom.

This assignment usually appears on the list as, Bring in an artifact and explain what it shows us about the life of the period. Generally, my students come up with a lot of interesting material. Some of it is fragile and valuable, some nearly indestructible. I have a closet I can lock, so artifacts usually go in there for safekeeping during the school day. Just to give you an idea of the possibilities: I have had a barbed wire collection, a 1920's dress, a World War I uniform, and a sample of Depression glass. Lisa did a very good job with the Depression glass. She set it out and explained what it was. Then she talked about the Depression and how these items helped people forget their troubles. It was a fine example of using an artifact to help people understand.

If the item is not fragile, handing it around is a good idea. Students have a lot more regard for the nineteenth-century housewife after hefting her old-fashioned sadiron. Those things are heavy. There is one other benefit to artifact reports. Holding an item in their hands while they give a report relaxes many students. They become less self-conscious and concentrate on communicating. After all, when you are talking about arrowheads, you want everybody to get the point.

Interpretive Reading

My students have a typical reaction to interpretive reading the first time they see it on the assignment sheet: What's that? I explain that interpretive reading is reading something aloud with meaning and expression. An interpretive reading usually has an introduction to set the material in context, but the interpretation is mostly done through the reader's voice, facial expressions, and movements. I allow two or three students to do interpretive readings as a group, but they more often do solo performances.

This assignment stresses communication skills and can expose students to primary source material. Students select and edit their material; then they then perfect the delivery. It approaches acting, except without costumes.

I bypass a lot of obvious material. I do not assign the Declaration of Independence, Washington's farewell address, Lincoln's Gettysburg address, or any of the other great American speeches, including Kennedy 's inaugural and Martin Luther King's "I Have a Dream" speeches. Instead, I use a relatively short list of books—not all great in the literary sense, but influential. I always include Upton Sinclair's *The Jungle* and Sinclair Lewis's *Babbitt. The Jungle* is not a particularly demanding book. It's simply a matter of finding the good parts and reading clearly. *The Jungle* is the book that pushed the federal government to take up meat inspection. An interpretive reading just before lunch of the selection on making sausage has undoubtedly caused a number of students to take just a salad.

Babbitt, a satire on a prosperous businessman of the 1920's, requires skillful analysis. It is a complex book that students can approach from various angles. They can find selections that describe physical aspects of the time: houses, cars, and clothing. Or, they can choose material that shows a businessman's attitude toward contemporary developments, like labor unions, changing morals, and prohibition. Or, they can find passages that reveal the author's attitude toward American society. My students almost always take the first option, often the second, and almost never the third.

The interpretive reading does not have to be selections from a complete book. If you have an anthology of documents, you could assign a talented student to prepare a document to read aloud. A good reader adds meaning to the document just by the way he or she reads it. That is what interpretive reading is all about.

Plays

Plays are fun, and they can add significantly to an American history class. There are two choices: a work that dates from the period or an imaginary role play set in the period. I'll give you my favorite example of each. The best thing about plays from the students' point of view is that a group of them can work on a play, and the audience always enjoys seeing their peers in costumes and makeup. The mechanics of getting people into costume are a little tricky in the average school. When people are excused to go to the restroom and they come back down the hall wearing a beard or old-fashioned dress, it can be a little disruptive, but it adds to the enthusiasm.

Usually my students elect to do part of a play. Only one class has ever done a full play. It was a lot of work and a lot of fun, but the class that did it was exceptional—I do not expect to see their like again. Usually, a group of two or three students signs up to do selections from a play.

I can't imagine teaching about the Civil War without including *Uncle Tom's Cabin*. *Uncle Tom's Cabin* was a novel, not a play, but it was adapted for the stage in 1852. It is clearly a very influential work; and once the students see some of the important scenes dramatized, they begin to understand the emotions of the time. From an acting standpoint, it is not very demanding. It is high melodrama, so it doesn't take an Oscar-level performance to get the point across.

Costume can be as simple as a kerchief tied around the head, an old-fashioned long dress or an old suit. Our student productions are presented in a regular classroom without raised stage or special lights, but the magic of the theater is there.

Role playing becomes, in effect, a play written by the students. In some cases they improvise rather than write it down, but they have to do research on a person in history and develop an interview. Usually I can suggest a book or another source that reveals a lot about the person. In role playing, the students sometimes ignore costumes and makeup, and sometimes they are very enthusiastic about them.

My favorite role-playing project is an Eastern reporter interviewing a mountain man. It is popular with students, usually selected every year. In the best version to date the student wore moccasins, handsewn wool pants, flannel shirt, and a broadbrimmed leather hat. He carried a "possibles" bag packed with equipment, a tomahawk, a pistol, and a rifle. For most of the period, he talked about his equipment and his life style. In this case the student played a mountain man whose hobby was target shooting with a muzzle-loading rifle. In another presentation, the students doing this project read the autobiography of a mountain man, and the reporter fed questions to the mountain man, whose answers were quotations from the autobiography.

Once, two students became fascinated with J.P. Morgan. It started with the usual assignment: interview him about his business and his wealth. As the year progressed, the students requested a special project to interview J.P. Morgan about developments in later units, like the Progressive Movement. They became so fond of the interview format that the next year they even talked their government teacher into letting them do it.

Tom and Gary, two of my favorite students, took historical role playing to its extreme. They ignored my list to page through the text looking for an interesting prospect for an interview or ask me to suggest one. Next they talked another student into playing the role. Then, they interviewed their victim in the best tradition of Sixty Minutes. "How do you explain the $100,000 you received? What were the charges against you in this court case? What were you doing with the Secretary of the Treasury's wife while he was out of town?" They were an otherwise likable pair, so they got away with it and retained their social standing, but as the year went on, they had more trouble finding people willing to submit to an interview.

A TECHNOLOGY NOTE

What follows can serve as a historical benchmark: It reflects computer technology as we applied it in the early 1980's. The assignments are still valid, but the technology is history. Assignments that were done on a computer then can now be done on a programmable calculator. The next chapter deals with some computer projects for the late 1990's and beyond.

We have an odd attitude toward machines. Students interested in cars are steered into auto mechanics classes and treated as nonacademic. Students interested in computers are considered academically gifted. Yet, cars and computers are both machines.

These projects were created for students who were computer buffs—not the video game players, but students who liked to solve problems with computers. They were also for teachers who had little or no experience with computers—me, for example. The problem facing me was to develop some projects in American history that used a computer. The solution was to mix in historical data and somehow have the computer process it. I'll start with the most complex project and move to some simpler ideas.

Chan was a special boy. He had skipped ahead of his own class and was a top student and a computer whiz. The obvious thing to do was to tap into his computer interest and challenge him. Fortunately, I had just read Dupuy's *Numbers Prediction and War,* a dense book in which the author develops a formula for predicting the outcome of battles. He runs a few sample battles through his system, including Waterloo and Gettysburg. My assignment for Chan was to select a battle in the American Revolution and run it through Dupuy's formula. Did the formula work? Why or why not? This project required him to read the Dupuy book, gather statistical data on the battle—we chose Saratoga—and devise a program. Chan breezed through it and gave a report that was a model of clarity and completeness.

On a less demanding level, my student war gamers, who met on Saturday, did a similar project. We found a formula to determine the relative power of World War II naval ships. For several weeks they pored happily over references, pulling out statistics on armor, guns, speed, and so on. One student wrote a program to compute the tables. We lost that copy, so another student created a second program, now secure in my files.

Economics is one field where the computer comes into its own. One economic history class created a computer model for inflation: We could take the price of an item today, project a level of inflation, say 5 percent, and then calculate the price of the item in five, ten, or twenty years.

Politics also generates a lot of numbers, and manipulating numbers on computers—looking for correlations, sorting and selecting data to see trends—is fun for student computer buffs. Some questions to investigate: From 1945 to the present, what has been the average percentage of the population voting (Democratic, Republican) in the nation? In presidential elections of the past 20 years, how has the inflation rate correlated with reelection of incumbents? The unemployment rate? Every ten years, states redraw the boundaries of congressional districts using computers. If your students can run down voting figures and population figures—public information—they can work up their own congressional districts.

An interesting computer project can be made from some very simple components: a question, some numbers, and a program. The teacher can provide ideas for interesting questions and hints on where to find the numbers. The students can take it from there.

VIDEOS

There is no reason why oral reports have to be delivered live. With camcorders, students can easily produce short videos. Check around: Your school or your students or colleagues probably have the equipment. My high school even has a studio that produces a program for public-access cable TV. Most of the reports in this chapter could be taped. I recently saw a video done by two talented students who had staged scenes illustrating rights under the Constitution followed by commentary that explained the law in each scene. They produced the whole thing with a wonderful deadpan humor. You might want to offer the video option to your students.

A Contest

If you have a talented student who enjoys the challenge of competition and has ideas about life in America, there is a great contest. The American Legion sponsors an oratorical contest on a citizen's duty under the Constitution. The national level first prize is $18,000, and state-level first prize is $2,000. In my area local contests also offer prize money or a series EE bond. The one my students enter awards first, second, and third prizes. (Local awards are decided by the local post, so there is no guarantee of cash prizes.)

Money isn't the program's greatest strength. The Legionnaires treat speakers very well. It is a very pleasant experience for students. When a local contest winner advances to the next level of competition, the local post provides transportation and sends along a cheering section.

The speech should be eight to ten minutes long, and for higher levels of competition, it should be memorized. At local contests, speakers may use notes. Speakers can approach the topic personally as long as it relates to a citizen's duty under the Constitution. I have heard speeches on gun control, fighting crime, free speech, immigrants, and abortion. The most popular topic is the duty to vote.

A mixed panel of educators (English or speech teachers heavily represented), usually a minister and a lawyer, and often a Toastmasters member judges the contest. They may or may not be Legion members. After the judging, the panelists offer sympathetic and helpful critiques. It is truly an elegant contest. The bibliography has information on finding local contests.

Bibliography

Arnold, Sam. *Sam Arnold's Frying Pans West Cookbook.* Denver: The Fur Press, 1969. (This is a selection of recipes from a television series on Western cooking. If boiled moose nose or buffalo tongue are not for you, perhaps you will try some of the breads. Useful for Flavor of History reports.)

Barrett, S.M. *Geronimo: His Own Story.* New York: Ballantine Books, 1970. (This is the story the old chief told through an interpreter. Geronimo is not exactly careful with the truth. A possible interview.)

Benedict, Michael Les. *The Impeachment and Trial of Andrew Johnson.* New York: W.W. Norton, 1973. (This is an interesting example of a historian using a computer. The thesis of the book is a challenge to traditional ideas about the impeachment.)

Boyington, Gregory. *Baa Baa Black Sheep.* New York: Bantam, 1977. (A famous flyer, World War II hero, and alcoholic tells his story. A potential interview.)

Brown, Dee. *The Gentle Tamers: Women in the Old Wild West.* New York: Bantam Books, Inc., 1974.

Douglass, Frederick. *Narrative of the Life of Frederick Douglass an American Slave Written by Himself.* New York: New American Library, 1968. (A reprint of a famous abolitionist tract and a potential interview.)

Dupuy, T.N. *Numbers Prediction & War.* New York: Bobbs-Merrill, 1979. (The formula explained in this book is the basis for a very challenging report for the computer buff.)

Lewis, Sinclair. *Babbitt.* New York: New American Library, 1961. (A novel portraying a typical businessman in the 1920's. Interpretive reading material.)

Riordon, William L. *Plunkitt of Tammany Hall.* New York: E.P. Dutton, 1963. (An insider's views on machine politics and "honest graft." A potential interview.)

Stowe, Harriet Beecher. *Uncle Tom's Cabin or Life Among the Lowly*. Norwalk, CT: The Heritage Press, 1966. (This is the novel, useful for interpretive reading, but there is a play script, Uncle Tom's Cabin by George L. Aiken. It is available from Samuel French, Inc., 25 West 45th Street, New York, NY 10010-2751.)

Sinclair, Upton. *The Jungle*. New York: The Heritage Press, 1965. (A primary source on life of workers in the early twentieth century and on meat packing of that era. A great book for interpretive reading.)

Truman, Margaret. *First Ladies*. New York: Random House, 1995. (Margaret Truman has known many first ladies. She reviews the history of first ladies with wit and an insider's perspective.)

Twain, Mark. *Life on the Mississippi*. New York: Airmont Publishing Co., 1965. (The memoirs of a riverboat pilot from the golden age of the steamboat. Good for either an interview or an interpretive reading.)

_________. *Roughing It*. Avon, CT: The Heritage Press, 1972. (Not only a very funny book, but a source of first hand information about mining and mining camp life. Useful for an interpretive reading.)

Wingo, Josette D. *Mother Was a Gunner's Mate: World War II in the Waves*. Annapolis: Naval Institute Press, 1994. (A memoir of women in the Navy and civilian life in World War II.)

United States Department of Agriculture. *Selections from Aunt Sammy's Radio Recipes and USDA Favorites*. Washington: U.S. Government Printing Office, 1976. (This inexpensive pamphlet includes recipes from the 1920's and 1970's. It is another source for Flavor of History reports. Known also as *Home and Garden Bulletin* no. 215.)

Washington, Booker T. *Up from Slavery*. New York: Dell Publishing Co., 1965. (The autobiography of a famous black leader. His comments range from early childhood memories of slavery to the segregated society of the South in the 1890's. A potential interview.)

For Legion Oratory contact:

The Americanism and Children and Youth Division, The American Legion, P.O. Box 1055, Indianapolis, IN 46206.
Web site: www.legion.org

CHAPTER XVI

COMPUTERS

Computers are the wave of the future. They will revolutionize education, make textbooks obsolete, and allow individualized instruction. They are, in short, the hot new educational panacea. I have been teaching awhile, and I have seen a few other educational cure-alls. Television was the first, followed by programmed instruction, team teaching, the new social studies, and cooperative learning. I've probably skipped a few. Simulation was once a big item too. In each case they were ideas with some solid potential whose advocates made inflated claims for them well. When sober reality settled in, they remained a part of teachers' tool kit and are used selectively where they fit best.

So now the computer advocates are in full cry, and every teacher is expected to become computer literate. My first reaction was, "Who? Me? I don't even like television much, and now I have to work with computers? I can't even use all the functions on my digital watch!" But there is no resisting change. Fortunately, I have worked with some excellent librarians and picked the brains of a colleague who is a computer expert. Then I have read a lot and spent some time on the Internet. I have found some ideas about using computers that look pretty practical. I'll try to give some sample lesson plans.

RESEARCH

Actually, I was teaching with computers for over a year before I realized it. My classes routinely write two-page research papers. I would assign the paper and then arrange for a workday in the library. Some of the paper topics were current, so my students used current periodicals. Our librarians carefully oriented students to the card catalog and periodical index. I had to "help" with the process and so learned how to use the system. I recently learned that the system, called "Infotrac," is not just a catalog; it is also a CD ROM that prints out articles. I hadn't changed my assignments at all, and some of my students were doing them using a computer.

Something similar happened with my current events assignment. Recently, I have been asking my students to turn in a news clipping and a summary of 50 words or less on a topic I announce a week in advance. Our library has a box of newspapers that can be clipped. Soon the question came, Can we use the Web as a news source?

My answer was, "Sure, as long as you give me a copy of the story and a summary." Our school has a computer lab, and many families have home computers, so I have seen quite a few current events clippings that were in fact computer printouts.

We have now crossed the line into letting students use the Web. This can be messy. Let's start with a relatively mild problem and get more dramatic. Amy was a serious student, but her paper on missionaries was unrealistically positive about their present success. At least it seemed so to me. I checked her bibliography and talked with her. She had taken her information off the Internet. Apparently, the articles she used were reports by a missionary group who weren't about to report failure. It seemed to me that Amy should have been a bit more skeptical about her sources.

The fact is that anyone can put almost anything on the Internet. All it takes is a computer, a modem, and the computer skills. High school students can do it. College students can do it. Neo-Nazis who deny the Holocaust can do it. People who have some very weird ideas about sex and the pictures to illustrate them can do it. The biggest problem isn't pornography, though; it's junk. The cliche is that about 90 percent of the Internet is trash. How do you find the educationally useful material, and what do you do with it?

At this point teachers need some computer skills. Fortunately, the skills required are pretty simple. You need a computer with a modem and an Internet connection. Once these are available, you need to find somebody who knows how to use the Internet. My first lesson took about an hour. Actually, learning how to get on took less than half an hour. We spent the rest of the time reading the Civil War diaries that we found. The skills you need are minimal: All you do is type a few commands, use a mouse to point on certain icons and click to activate. Icons are little pictographs. A mouse is the little hand control that moves a cursor on the screen. My five-year-old grandson can use a mouse: You really don't need a class. All you need is a few minutes of instruction from somebody who knows how to use the Internet. After that, you can start wandering, or "surfing," on your own to get a feel for how it operates. You usually search by typing in words to generate a list and then pointing and clicking at things on the list until you find something you like. I recently typed in "Civil War" and got 100 sites on the topic. The machine had selected only the 100 most popular sites. There were lots more! Do you see a problem here?

Obviously, there is a need to get directly to useful sites. The solution is simple. Collect addresses for some good sites. There are many sources: A few are discussed in the next few pages, and the bibliography has web sites. Most organizations include their web address in their advertisements or on their letterheads.

Let's do an example using the Civil War. A site like "Lesson Plan: The Civil War" looks good. The address is http://www.smplanet.com/civilwar/civilwar.html. Once you've reached that site, you decide to look at the site "Outline of the Civil War." The address is http://www.cais.com/greatamericanhistory/gro2007.html. But you don't need to type it in. All you need to do is put the cursor on the word "Outline" and click. Later, if you want to go directly to the outline, you can just type the address.

Once on the outline, you decide to investigate the topic "Monitor vs. Virginia." It yields two addresses: http://www.cwc.lsu.edu/pictures/ironclad.html (a picture of the battle)and http://tqd.advanced.org8012944monitor.html (an essay entitled "The Duel between the Monitor and the Merrimack," by Nicole Preziosa). Who is Nicole Preziosa? There is no identification. The essay is factually accurate as far as I can tell. It is in the first person, but I doubt that she was there to participate in the battle. This might be a project for an advanced placement high school history course.

Sometimes you will find highly qualified sources. You can also find abundant material from unquestionable sources on the Web, for example, at Special Collections of the Library of Congress, Civil War Photograph Collection (http://leweb.locgov/spcoll/048-2.html). This source had six pictures taken during the Civil War, authentic primary source material that students can access in seconds.

You have probably noticed that the information retrieved above is relatively limited. The printout from each site would probably be only a page or two long. More extensive material is also available on the Internet. For example, the African American History site (ftp://ftp.msstate.edu/pub/docs/history/USA/Afro-Amer) has a long list of subtopics. One is "bookertwash." Click on that and you get the full text of *Up from Slavery.* Other listings yield *Narrative of the Life of Frederick Douglass an American Slave* and the text of Martin Luther King's "Letter from Birmingham Jail." There is also a long entry on the Black Caucus in Congress. I discovered all of this in one relatively brief visit to the site.

Once you have located an Internet source that you want to use, you have some decisions to make, taking into account your skill, your students' skills, and available computers. Planning how to use the material is like planning to use any good resource—book, magazine, or video. You could decide simply to download the material and print it out. At that point you are on familiar ground: Make the copies you need and go for it. One caution: Be careful about copyrights. Most Web material seems to be in the public domain, but not all is. Using a paper copy of Internet material, of course, means that students don't have the chance to use the computer themselves. As far as the students are concerned, a printout is just another piece of paper. Although I have read a lot about how much students are interested in computers, in my personal experience, some are very interested, but it's not universal. I offered one of my classes the option of working on computers and they actually turned it down.

If you want your students to work on computers, you can visit the site yourself in advance and download the material you want to use, storing it in your computer's built in memory or on a portable disk. Information is available when you want it, and you avoid certain problems, including students getting sidetracked on the Web and electronic traffic jams and glitches. Also, the computer you use does not have to be connected to the Web at all. Anything that can read your disk will work.

Assume that you want your students to read Martin Luther King, Jr.'s "Letter from Birmingham Jail." You go to the African American History Web site and then locate the document. (You can add a document's address to your "bookmarks" so your computer will go right to it at the click of a mouse.) You consider your computer resources and your class. Is a computer available for each student? Do all computers have an Internet connection? If so, you can simply have your students pull up the document on their computers and do the study guide on page 134.

Some experts say that teachers don't need a computer for each student. They argue that groups of students can work with computers. A class of 30 could form clusters of five students around six computers. I have no personal experience with such a setup, but presumably it combines teamwork and technology.

Computer Presentations

Some computer programs let students or teachers do presentations. Hypercard, Electronic Chisel, and other programs allow students to assemble information on a screen layout or screen. They can include graphics, some text, and sound. Students can build their layouts on the screen. These then become class presentations or projects. You and your students can also enliven your presentations with CD Rom's, a computer, and a monitor.

I was visiting a chemistry classroom at my school recently and noticed a computer keyboard on the teacher's desk and a screen mounted on the wall. What the teacher typed appeared on the screen—a high-tech blackboard. Computers have endless possibilities.

Bibliography

Headings, Michael D. *Teaching American History with the Internet.* Lancaster, PA: Classroom Connect, 1997.

__________. *Teaching the Civil War with the Internet.* Lancaster, PA: Classroom Connect, 1997.

(Both of these are cookbooks for teachers. They contain lesson plans completely laid out and handouts ready to go. Lessons are K–12. Web site is http://www.classroom.net and e-mail is connect@classroom.net)

King, Martin Luther, Jr. "Letter from Birmingham Jail," *Why We Can't Wait.* New York: Harper & Row Publishers, 1963.

Levine, John R, Carol Baroudi and Margaret Levine Young. *The Internet for Dummies,* 3rd ed. Chicago: IDG Books Worldwide, Inc., 1995.

(This was the only introductory computer book I could actually read.)

"Technology and Social Studies." *Social Education,* Vol. 61, No. 13 (March 1997).

Name ______________________________

Date ______________________________

"LETTER FROM BIRMINGHAM JAIL"

STUDY GUIDE

1. Why was Martin Luther King in jail in Birmingham?

2. Why was he writing this letter?

3. What were the "four basic steps" in a nonviolent campaign? How did they develop in Birmingham?

4. What was Martin Luther King, Jr.'s position about unjust laws?

5. Why was he disappointed with white moderates? With the white church?

6. What values was he using to support his position? Give examples.

7. What was his hope? Have we achieved it?

8. Write a letter back to Martin Luther King, Jr., telling him about our progress or lack of it since 1963.

CHAPTER XVII

DEBATE

This country was born because of a debate. Richard Henry Lee's resolution that "these United Colonies are, and of right ought to be, Independent States . . ." was presented to the Continental Congress, weighed in a parliamentary debate, and passed. Parliamentary debate ranging over an endless list of topics is still ongoing—in the constitutional convention, in Congress, and in state legislatures.

In 1787 the question in state conventions was, Shall we adopt the proposed constitution? Since the Constitution was passed, each law and amendment has been debated in Congress. Among the more famous congressional debates is the Webster–Hayne debate.

Election campaigns have produced debates, sometimes of tremendous impact. How great was the influence of the Lincoln–Douglas debates? They are hardly of the same calibre, but we also have television-era presidential debates. It could even be argued that the history of America is the history of a series of debates. That would offend some historians and would, in turn, lead to a debate. Historians do almost as much debating about American history as there is debate in American history. Was slavery the major cause of the Civil War? Was the Constitution shaped by economic interests of the founding fathers? Is America really a melting pot? The list goes on and on. Americans today are debating NATO expansion, the president's budget, and school choice. On the state level in Iowa, we are debating hog-lot regulation and a ban on nude dancing. I don't pretend this list is complete.

Several years ago I pilot tested a unit in sociology about what makes a country democratic. One of the main ideas was that democratic government takes strong root in societies that include many small organizations that are centers of discussion and debate.

Whatever your goals in teaching American history—facts, structure of the discipline, citizenship, skills—debate can contribute to any of them, and at minimal cost. Debate can strengthen library skills by encouraging research. It requires students to read carefully and take systematic notes, analyze issues carefully, and exercise speaking skills. All of these benefits come from an activity that calls only for reading material, pen or pencil, note cards, and paper. A school could buy all the materials needed to have classroom debates, books included, for about half the cost of one computer. If the school has a good library already, the cost of debate drops to a few cents per student.

The next few pages explain how to set up a debate, common student mistakes and how to deal with them, and where to get materials for debate. They also explore some sample debates from American history and some from my own classes and suggest a list of debate topics.

WEBSTER–HAYNE

The Webster–Hayne debate was a great sensation in its day and has echoed through American history ever since. It also provides a model for one way to organize a debate. Webster and Hayne were both senators. Their debate took place on the floor of the Senate, and the format was parliamentary debate. A resolution about states' rights to nullify federal law was before the

Senate, and speakers were recognized to speak either for resolution or against it. Although a number of other senators spoke on the topic only Webster and Hayne are remembered. Hayne spoke first and Webster replied the next day. For parts of the next two days, Hayne answered Webster; then Webster took the floor to deliver what is often called the greatest speech in American history. It ran several hours over two days. The Senate was packed with spectators, and the newspapers reported the speeches to eager readers. For years afterward, students in the North were required to memorize parts of Webster's reply to Hayne.

If you adjust a few details, your class can use the same format: The principles still work. You need to select a topic, perhaps one from the list included in this chapter. (In the 1830 debate the topic was limiting the sale of public land). Then, form teams: You could poll the class assign them to sides according to their convictions, or you could use an arbitrary process. Any even division of the class will do. Give students some hints about where to do research and allow them some time to get materials. I prefer that students do research during study halls, evenings, and weekends. My students often hint that a more humane teacher would allow class time for research.

On the day of the debate, the presiding officer calls the meeting to order and announces the topic. After that, it's a matter of recognizing speakers. The two sides take turns. The presiding officer recognizes speakers from the two sides alternately and should try to let every student in the class have a chance to speak. Thus, the debate involves the whole class. When you feel debate is exhausted, call for a vote on the resolution. The presiding officer can be either a student or the teacher. You may want to limit speeches to, say, five minutes. My students give very short speeches, like thirty seconds, so I have never needed a time limit. Following strict parliamentary procedure is an elegant touch, but the substance of the debate is just as good if you ignore technicalities and just recognize speakers from alternating sides. Grading can be a problem. I keep a running set of notes on speeches and award points for each. At the end of the debate, I total up all points and make up a curve, just as if it were a test. In some cases, I have not even bothered to grade the debate because the desire to win is enough to motivate students to try hard.

This format works well for most classes, even those with little debating experience, and the teacher does not have to be an expert. I have used this format for debates for over 20 years, the first time as a student teacher at the Iowa Braille and Sight Saving School. All the students were legally blind, so I had to read the research material to them. Their topic was, "Resolved: That the United States should adopt a national health program like Britain's." None of the students had debated before, but they attacked the question with vigor and enthusiasm, and it was a good debate.

Lincoln–Douglas

Parliamentary debate may be the most common form of debate in America, as it is in my classroom, but there are other useful forms. The Lincoln–Douglas debates provided not only a set of historic quotations, but also a classic debate format. When Lincoln and Douglas squared off, their nominal topics for debate related to slavery, but what they were really arguing was, Resolved: That A. Lincoln should replace S.A. Douglas in the Senate of the United States. The two men agreed to seven debates in towns throughout Illinois. Debates were major events, with parades, banquets, cannon fire, and crowds of up to 20 thousand. Debates were held outdoors and lasted for three hours. The first speaker spoke for an hour, his opponent responded for an hour and a half, and then the first speaker gave a half-hour rebuttal. There were no loud-speaker systems in those days, so speakers had to be powerful physical specimens. Imagine speaking for ninety minutes loud and clear enough for 20 thousand people to hear every word! That's something to mention the first time a student starts to mumble and the class can't hear. Better yet, it's something to mention before they start: A good speaker is one who can be heard.

The Lincoln–Douglas debate format is the basis for academic debating competitions. Modern Lincoln–Douglas debate includes cross-examination and uses questions about values.

The affirmative speaker is allowed six minutes and then is cross-examined for three minutes by the negative speaker. The negative speaker is allowed seven minutes and then is cross-examined for three minutes by the affirmative speaker. A four-minute affirmative rebuttal is followed by a six-minute negative rebuttal; then the affirmative speaker delivers a final three-minute rebuttal to end the debate. You can adjust these time limits to suit the students and the topic.

Debate topics that concern values are more abstract than those that concern policy. In competition, it is understood that debaters will establish criteria to support their position. For example, in debating a topic like "Anarchy is better than tyranny," the affirmative side could support John Locke's theories on government and rights, while the negative could support Thomas Hobbes. How many students do you have who are familiar with seventeenth-century philosophers? In the classroom, of course, this particular format is not required to debate value-oriented topics. A few sample value-oriented topics are included on the list in this chapter, but be advised this is a very hard kind of debating for neophytes. If your school has a debating team, they will be familiar with this form.

The twentieth century has not given us many orators to match up to Webster and Hayne or Lincoln and Douglas. It has also given us the televised presidential debates, with larger audiences and shorter speeches. The Kennedy–Nixon debates provided us with another useful format. In their first debate, each speaker made an eight-minute opening statement, Kennedy first. A series of questions to the candidates from a panel of television reporters followed, then a three-minute concluding statement from each. Historically, this debate is usually remembered because Nixon looked tired and almost ill. (This is another point to make to students. A good debater looks confident and vigorous.) For class, the format could become eight minutes for the first speaker, eight minutes for the second speaker, four minutes of questions from the class, then three minutes of rebuttal from the second speaker followed by three minutes of rebuttal from the first speaker. Again, time limits should be adjusted to match the speakers' skills. All time limits should be understood to be maximums, with no penalty for quitting early. In fact, quitting early is preferable to wandering on and making a foolish mistake.

While this doesn't seem to be based on any historical debate format, traditional academic debates on policy topics are contested in teams of two. On the summary chart of debate formats on page 141 there is one for two-person teams. On page 137 Ten Guidelines for Debaters summarizes the main ideas of this chapter. You may make copies of the guidelines for your students to use.

Preparation

To set up a classroom debate, first pick a debate format so you can decide which students will participate. Will you let the whole class participate in a Webster–Hayne type debate or select your two best to debate in the Lincoln–Douglas style?

The topic can be any question that can be answered with yes or no: Was Slavery the principal cause of the Civil War? Should the United States adopt an equal rights amendment? The topic can also be stated as a resolution, "Resolved: That the United States should adopt an amendment to the Constitution of the United States relative to equal rights for men and women." As long as there are two sides, an affirmative and a negative, it's a debate topic. The affirmative argues that it is true, or should be adopted, depending on how the topic is stated. The negative argues it is not true or should not be adopted.

Every topic calls for some analysis. Note that the Civil War topic on the list says "principal" cause, not "only" cause. A shrewd affirmative would undoubtedly argue that slavery may not have been the only cause, but it was the most important cause. "Principal" means "first," as in "first in importance." The affirmative side can admit that there may have been other causes but argue that slavery was the principal cause and so prevent the negative side from merely listing other causes. They have to refute the affirmative's arguments that support slavery as a first cause or else suggest and support an alternative first cause. The second topic in the preceding example looks simple. It seems to call for what is usually called the Equal Rights Amendment to be added to the Constitution. There was only one text proposed for the amendment, but the topic says "an," not "the." Could it be that an affirmative side could draft their own version of the Equal Rights Amendment? Probably, but they would have to be careful that it is an amendment "relative to equal rights for men and women" or the negative could argue that they are not dealing with the topic. Defining the topic in a debate can be very important.

A value topic needs to be analyzed differently. The affirmative side is not obliged to advocate any plan of action. (In competition debates on value topics, plans are actually ruled out.) The focus is on ideas and ideals, not action: Are values named in the topic? What are we evaluating? Is the evaluation relative or absolute? In the topic "Anarchy is better than tyranny," two ideas are being compared. The phrase "better than" does not necessarily spell out a value position. Does it mean "always better than" or "usually better than"? Sometimes the affirmative and the negative sides will not agree on definitions of important terms like "anarchy" and "tyranny."

If you want to use current topics, it's easy just to write your own. Often the headlines will write them for you. At this writing (1998), the following topics look lively: Should the Internal Revenue Service be abolished? Should workers be subject to random drug tests by employers? Should NATO be expanded to include Eastern European countries? Should the U.S. end affirmative action? While a topic may be hot in the fall of 1998, it probably will cool off quickly. But new issues keep coming up.

While the above topics are relatively current, they don't require students to relate the issues to recurring questions in U.S. history. During the U.S. Bicentennial in 1976, a group of experts developed a series of topics designed to explore recurring questions. They provide an excellent framework for analyzing current issues in a historical context. Here are some sample ideas for using the topics to relate past and present. The analysis provided is not from the original booklet and is not exhaustive, but it is based on lots of experience with debate and U.S. history.

1. Resolved: that the "melting pot" metaphor is a historical fiction. Have the descendants of immigrants in the nineteenth and twentieth centuries blended into the melting pot? Is the United States truly a multi-cultured society?

2. Resolved: that urbanization has lowered the quality of American life. How does life in the cities today compare with life in the U.S. in 1789, 1860, or 1920? Has pollution gotten worse? Is there really more violence? Is the modern drug problem worse than the patent medicines of an earlier day?

3. Resolved: that extremism in the defense of liberty is no vice. Who is an extremist? Do should we condemn the minutemen of 1776, the abolitionists before the Civil War, or the Civil Rights movement of the 1960's? How does the militia movement of the 1990's compare to the older groups? Does a teenager who violates a skateboarding law qualify as protesting an unjust law? Would refusal to register for the draft in 1970 or 1998 be justified?

4. Resolved: that American political parties have been dominated by socio-economic elites. Were the founding fathers an elite? Did the wealthy dominate political parties in the nineteenth century? Does current campaign finance law let the rich buy officials? Are the Kennedys, Tafts, Adams, and other families dynasties? How important has the common voter been?

5. Resolved: that America has significantly overemphasized social welfare at the expense of individual rights. Has America emphasized fighting crime at the expense of the rights of the accused? Have we conducted unjustified witch hunts for communists? Have schools' concerns about drugs and weapons led to unjustified drug tests and locker searches?

6. Resolved: that government policy toward the American economy has benefited consumers at the expense of producers. Were the nineteenth century laws regulating railroads damaging to the railroads? Historically, has our trade policy favored consumers or producers? Who does the North American Free Trade Agreement (NAFTA) help more, consumers or producers? Do our current consumer protection laws excessively restrict producers?

7. Resolved: that American foreign policy has wrongly violated the basic principles of the Declaration of Independence. What are the basic principles of the Declaration of Independence? Has our aid to dictators like Joseph Stalin, Manuel Noriega, Ferdinand Marcos, and Saddam Hussein been wrong? Or was each case justified by a greater evil threatening us?

8. Resolved: that American public education has emphasized socialization to the detriment of intellectual development. Has the effort to integrate public schools hurt intellectual development? Have efforts to increase self-esteem hurt intellectual development? Has bilingual education hurt intellectual development?

9. Resolved: that the history of America has been the history of a decline in the importance of the individual. Has the growth of huge corporations reduced employees' importance as individuals? Has the widespread use of computers in American society reduced the importance of individuals? Is the individual more important today than in 1776?

Questions are the only starting point. Questions don't change much over time, but the answers do. The quality of a debate will depend heavily on research.

Research

Usually research goes hand-in-hand with analyzing the topic. A debater needs to look up the historic arguments about causes of the Civil War or find out why the states failed to ratify the ERA to figure out which arguments became important. Most debates are won or lost in libraries. The team that has done the best research usually wins. Classroom debaters usually underestimate the amount of research it takes to prepare for a debate. Even a relatively short debate can require twenty or more evidence cards. Academic debaters routinely build files of several hundred cards. The debate evidence card is an independent unit. It contains a notation of the source, fully identified, and the source's qualifications followed by a full, accurate quotation from the source. Any attempt to falsify evidence or change meaning by revising wording is a serious violation of debate ethics. Evidence cards are sometimes organized in a file box so they can be easily pulled and ordered for use in speeches. It is more common for debaters to prepare pages of briefs with arguments outlined and evidence included. Top debate teams pull around little carts stocked full of briefs.

The hard part is finding the sources in the first place. You should be sure before you select a topic that material is available. If your students are like mine, they will probably need some help in locating it. Sometimes more direct action is necessary.

"Mr. Scott, this is Leonard."

The voice on the phone was familiar. I knew Leonard was debating tomorrow, so I half-knew the answer to the question I asked. "What's the problem, Leonard?"

"I'm down here at the public library, and I can't find anything on my side in the debate."

"That's odd, Leonard. There should be quite a bit on it. Are you sure you have looked at all possible sources?"

"I've looked everywhere, Mr. Scott."

"Okay, let's meet down there, and I'll try to locate something for you."

Complications arose. Leonard had a job, so he had to leave the library and would be unable to return until late in the evening, when I had a meeting. We finally agreed that I would go down, find something for him, and give him a quick phone call. He could then pick up the sources.

For some reason, Leonard was doing the hard part of his side. The topic was, "Resolved: That the Southern slave was as well off as the Northern working man." Leonard was on the negative. It would seem like an easy research job. All the negative had to do was to research all the descriptions of the horrors of slavery, and the debate should be theirs. Actually Leonard's team had analyzed better than that. They realized that they also needed to defend the idea that Northern workers were not too badly treated. Leonard was working on this aspect while his partner was responsible for the material on slavery. Splitting up the work was a good idea, but Leonard's part was hard to research. My first few minutes were frustrating. Pre-Civil War working conditions are not the most popular topic in labor history. Finally, the catalogs and indexes began to yield very good evidence. I found reports of clean factories, short hours, and even one case where workers hired a boy to read to them as they worked. I passed the sources on to Leonard.

The next day I settled back to watch the show. With the stuff I had dug up, Leonard should win hands down. Alas, it was not to be! The affirmative had done superficial research, but they were superbly organized and excellent at replying to negative attacks. Leonard's partner had hardly anything to say. I still had great hope for Leonard, but his speech made it clear that he had used only about a third of what I had uncovered, and far from the best third. Frustration!

Actually, both teams were doing several things right: analysis, organization, and speech delivery, for example. But they were also doing one important thing wrong. They had not done enough research. A debater who starts research the night before the debate is in real trouble. (You, of course, will be wise enough to start your students on research at least a week before the debate. Perhaps you will also check their card files the day before the debate.)

Evidence cards cause the most common mistake that beginners make in their presentations—the "I have a quote" syndrome. New debaters feel that just reading evidence wins the debate. They forget that the argument must first be clearly stated and the evidence must be carefully related to the argument. In the worst cases, new debaters do nothing but read evidence without adding a word about what it means. Really polished, experienced debaters weave evidence into the debate smoothly and relate it clearly to the argument.

Sample Debate

In the following hypothetical example, I have tried to show a relatively skillful presentation of evidence. This is only a fragment of a debate, but it should give some idea of how an argument develops and evidence is presented. This debate is on the topic "Resolved: that American foreign policy has wrongly violated the basic principles of the Declaration of Independence." It is set up for an academic debate with two-person teams, although arguments could develop in the same way regardless of the debate format.

The first affirmative speaker opens the debate by introducing the topic and defining key words in the topic. Next, the first affirmative begins the affirmative case.

"The affirmative is going to argue that the United States has wrongly violated basic principles of the Declaration of Independence. First we will show you what those principles are; then we will show you that our current policy toward China violates them.

"Let's look at the Declaration of Independence in our textbook. The second paragraph says, 'We hold these truths to be self-evident, that all men are created equal, that they are endowed by their Creator with certain unalienable Rights, that among these are Life, Liberty, and the pursuit of Happiness.'

"The affirmative team will argue from this evidence that life, liberty, and the pursuit of happiness are rights, and that denying those rights violates the basic principles of the Declaration of Independence.

"Now let's turn to our current policy toward China. China today is receiving most favored nation trade status. China is also violating basic human rights, and it's getting worse. As Mike Jendrzejczyk, Washington Director of Human Rights Watch/Asia, tells us in *Congressional Digest* of August-September, 1995, page 212, 'We have documented the decline of human rights in China and Tibet since President Clinton's MFN (sic) decision last May [1994]. The worsening human rights conditions are well described in the State Department's own human rights country report for 1994, issued February 1, 1995.'

"This evidence clearly shows that since gaining MFN status the human rights situation in China is getting worse. This is an example of American foreign policy wrongly violating basic principles of the Declaration of Independence. We are favoring an oppressive, tyrannical government."

The speaker makes a few more points and then finishes. In the "transcript" you may have noticed that the affirmative speaker didn't use "quote" and "end quote." Most debaters use voice and body cues to signal the start and end of a quotation. The debater will hold the evidence up prominently when reading it and snap it down dramatically at the end. The voice takes on a distinct tone that says, "I am reading evidence."

While the first affirmative debater has been speaking, the negative side has been taking careful notes. Now, the first negative speaker has to refute the first affirmative. Refutation is a demanding skill. It calls for carefully recording the opponent's argument and then beating it in the next speech. For the sake of clarity, the negative debater restates the argument, then makes the counter argument, presents evidence to support it, relates the evidence to the counter argument and demonstrates how this affects the debate. Let's join the first negative in progress.

"The affirmative has told you about human rights problems in China and has tried to argue that China violates basic principles of the Declaration of Independence. We will argue that they have yet to prove this. First, they have yet to show that human rights abuses in China impact on the same rights named in the Declaration. Second, we will argue that our current China policy is the best way to improve human rights in China. As the Honorable Max Baucus, U.S. Senator from Montana, said in a speech reprinted in *Congressional Digest*, August-September, 1995, page 209, 'It is just over a year since President clinton renewed MFN (sic) status for good in 1994. Since then we have adopted a policy that works. Our strategy is correct: to bring China in line with international standards of behavior on peace and security, environmental protection, human rights and trade.'

"Senator Baucus is telling us that our policy works to bring China in line on human rights, among other things. Thus it could not be in violation of basic principles of the Declaration of Independence." The negative speaker then moves on to other issues.

The second affirmative speaker is busily taking notes on all this, and if this is a capable team, has an answer ready before the argument is made. The second affirmative reply could run like this: My partner showed you human rights abuses in China after we started MFN status for them. The negative argues that our policy is working. I will reply in two ways. First, my partner's evidence said that after MFN status was granted, human rights in China continued to decline. Second, more recent evidence indicates that there is no basic improvement. Consider this from David Shambauch of George Washington University in *Current History*, September 1997, page 244.

'Administrative opponents have no shortage of complaints about the Chinese regime's behavior: its abysmal human rights record (a point conceded in the annual 1997 State Department human rights report) and treatment of dissidents; its repression of religious right; its harsh birth control policies . . .'

"Notice that this evidence is more recent than the negative evidence. Thus it shows that our policy has not changed China's behavior. The Chinese government continues to violate the basic principles of the Declaration of Independence and our policy supports them. The negative wins the issue."

The second affirmative neglected to reply to the argument about showing that the same human rights are involved in China as are named in the Declaration of Independence. Perhaps the negative will pay for this in later speeches. The issue may bounce back and forth in the rebuttal speeches as well, but you now have some examples of how to weave evidence into a speech and how to refute.

Cross-Examination

Two of the debate formats I have included call for cross-examination. Cross-examination in debate is much like that in courtrooms. The questioner is trying to get the witness, in this case the other debater, to admit something that will help the questioner win the argument. It's not as easy as it looks. It's amazing how often students forget these guidelines. Ask questions. Don't make a speech. Plan your questions in advance. Don't rely on hidden inspiration. Use a series of questions to lead your opponent toward your position. Be polite. It turns listeners off when you bully. Use a simple "thank you" when you have heard enough. The questioner controls cross-examination time and can stop the answer at any time. If you don't understand something, ask about it. This prevents nasty surprises later in the debate.

If the opponent agrees with a key idea on your side, nail that down in questioning. Remember, all this is to help you win the debate, so use answers in your speech. (These guidelines are summarized in a cross-examination handout on page 138.)

Who Won?

One of the most frequently asked questions about debate is how you know who won. The best answer is that the speaker who persuaded the target audience won. In high school competition, a single critic-judge or a panel decides the winner. In politics the voters decide. In your class, the students can decide. I like to use a shift-of-opinion ballot. Before the debate the moderator announces the topic and asks the audience to vote on it: for, against, or neutral. The vote can be by a show of hands or by paper ballot. After the debate the moderator asks the audience to vote again. The side that has shifted the opinion of the audience toward their side wins. For example, if the first vote is 10 affirmative, 8 negative, and 5 neutral, and the vote after the debate is 11 affirmative, 3 negative, and 9 neutral, then affirmative won. This is a good way to involve all students in the debate. It's even better if they have to write a paragraph explaining their decision.

You now have the basic tools, a set of guidelines for debaters, and a variety of debate formats. On the following pages are a list of topics and a list of sources on debate. Now all you need to do is add students and enthusiasm. Who know, maybe some exuberant candidate who has just won a big election will someday tell the press, "I won this election in the debates. It all started in American history class years ago. We had this teacher . . ."

Bibliography

Huseman, Richard C., and James I. Luck, eds. *Bicentennial Youth Debates Issue Analysis*. Washington: Speech Communication Association, 1975. (Contains the series of debate topics reprinted below and aids for preparing to debate them.)

Congressional Digest consists of pro and con viewpoints on issues before Congress. It is a helpful source for ideas for debate topics as well as evidence to support arguments. Available from: Congressional Digest Corp., 3231 P Street, N.W., Washington, DC 20007.

Name ________________________________

Date ________________________________

TEN GUIDELINES FOR DEBATERS

1. Treat your opponent with respect and courtesy. You disagree, but this is no excuse for sarcasm or hatred.
2. Remember, your audience must hear and understand you. Speak in a loud, clear voice, and explain ideas so that they can be understood.
3. You will be judged in part by physical appearance. Dress neatly. Act confident. Use good posture. Look the audience in the eye.
4. Listen carefully to your opponent. Take notes on his or her speech.
5. When you refute one of your opponent's ideas,

   (a) Tell the audience the idea you are attacking;

   (b) Make your argument against;

   (c) Give the evidence for your position;

   (d) Show how this evidence defeats the opponent's argument;

   (e) Show how it affects the whole debate.
6. The whole point of reading evidence is to make an argument stronger. When you read evidence,

   (a) State your argument;

   (b) Make a transition from the argument to the evidence;

   (c) Read the evidence, accurately naming the source;

   (d) Tie the evidence in with the argument.
7. Organize your speech so that it has an interesting opening, a clear body, and a strong ending.
8. When you run out of material, summarize, and sit down.
9. Be sure you understand the topic. Think about each word. If in doubt, ask the teacher.
10. Research will win 75–80 percent of debates. Research the topic as thoroughly as possible. The debater who knows the topic best almost always wins. (But if you fail in any of the above, you can still lose it.)

Name ______________________________

Date ______________________________

GUIDELINES FOR CROSS-EXAMINATION

1. Ask questions.
2. Plan questions in advance.
3. Be polite.
4. The questioner controls the time and can stop any answer.
5. Use questions to clarify ideas.
6. Use questions to set up arguments.
7. Use the answers in your speech to help your argument.

SAMPLE EVIDENCE CARD

China—Long-term Trends Positive

"Should the Untied States Government change its foreign policy toward the People's Republic of China? Con"

Speech by Max Baucus, U.S. Senator, Montana. *Congressional Digest,* August-September 1995, p. 211.

"We should also recognize that long-term trends are positive. To cite a few examples, China's first Koran recital competition since 1949 shows some progress on freedom of religion."

DEBATE TOPICS

POLICY TOPICS

The following deal with public policy—that is, they ask, What should we do? In debating a policy topic, the affirmative must propose a policy and argue for its adoption.

1. Should the United States adopt an amendment to the Constitution of the United States relative to equal rights for men and women?
2. Should the United States enact a constitutional amendment banning abortion?
3. Should the United States abolish the electoral college?
4. Should the "exclusionary rule" of evidence be dropped?
5. Should the United States enact a constitutional amendment to permit voluntary prayer in the public schools?
6. Should the United States ban the sale, possession, and use of handguns?
7. Should the United States enact a constitutional amendment requiring a balanced federal budget?
8. Should the United States adopt a policy of nonintervention in Latin America?

VALUE TOPICS

1. Resolved: That anarchy is better than tyranny.
2. Resolved: That when in conflict, individual rights should take precedence over the will of the majority.
3. Resolved: That there is a moral responsibility to disobey unjust laws.

HISTORICAL INTERPRETATION TOPICS

The following topics deal with making judgments about facts and values in American history.

1. Has the frontier been a significant factor in shaping American character?
2. Were the Articles of Confederation a workable framework for government?
3. Was the Mexican War a case of American aggression?
4. Was the Southern slave as well off as the Northern worker?
5. Was slavery the primary cause of the Civil War?

Name ______________________________

Date ______________________________

HISTORICAL INTERPRETATION TOPICS (continued)

6. Was congressional Reconstruction excessively severe?
7. Should the United States have dropped the atomic bomb on Japan?
8. Should the United States have fought in the Vietnam War?
9. Was Richard Nixon guilty of an impeachable offense?
10. Did Ronald Reagan's policies win the Cold War?
11. Is the foreign intervention policy of the United States based on pursuing economic interests or defending democratic principles?
12. Has the United States done enough to protect the environment?
13. Has participation in the United Nations advanced American interests?

U.S. 1976 BICENTENNIAL YOUTH DEBATES

The following topics were drafted for use in a series of debates which were part of the U.S. Bicentennial celebration in 1976. They are reprinted here with permission from the Speech Communication Association.

1. Resolved: That the "melting pot" metaphor is historical fiction.
2. Resolved: That urbanization has lowered the quality of American life.
3. Resolved: That extremism in the defense of liberty is no vice.
4. Resolved: That American political parties have been dominated by socioeconomic elites.
5. Resolved: That American foreign policy has wrongly violated the basic principles of the Declaration of Independence.
6. Resolved: That American public education has emphasized socialization to the detriment of intellectual development.

TABLE OF DEBATE FORMATS

ADAPTED FOR CLASSROOM

Webster–Haynes*

Parliamentary: teams of any size; speaking time unlimited; time for debate may be limited in advance. (One class period)

Lincoln–Douglas

Affirmative	6 minutes
Cross-examination	3 minutes
Negative	7 minutes
Cross-examination	3 minutes
Affirmative rebuttal	4 minutes
Negative rebuttal	6 minutes
Affirmative rebuttal	3 minutes

Kennedy–Nixon*

First speaker	8 minutes
Second speaker	8 minutes
Questions from class	4 minutes
Second speaker's rebuttal	3 minutes
First speaker's rebuttal	3 minutes

Academic Debate**

First affirmative	8 minutes
Cross-examination	3 minutes
First negative	8 minutes
Cross-examination	3 minutes
Second affirmative	8 minutes
Cross-examination	3 minutes
Second negative	8 minutes
Cross-examination	3 minutes
First negative rebuttal	5 minutes
First affirmative rebuttal	5 minutes
Second negative rebuttal	5 minutes
Second affirmative rebuttal	5 minutes

---

* Shortened and simplified for classroom use.

** Speeches should probably be shortened for classroom use.

CHAPTER XVIII

CLASSROOM TRIALS

Fads come and go, in education as in clothing. I haven't done a classroom trial in American history since miniskirts were in fashion. A few miniskirts are around again, and I've been reading about a movement to use mock trials as a part of law-related education. Maybe it's time to de-mothball the material on classroom trials and update it with information on the new movement.

Classroom trials can be valuable in at least two ways: by focusing attention on conflicting views of an important personality in American history and by offering students a chance to apply the Bill of Rights in specific cases. A lot of trial procedure stems from due process and other fine phrases out of our Bill of Rights. My experience is with trials of historic figures. Hypothetical cases designed to illustrate the legal process can be just as valuable as those focusing on people in history; I just haven't tried that. A trial requires about a week to set up, prepare, and run.

GETTING THE CLASS TO TRY A CASE

Usually, I start with a student who has an idea and is convinced it is right; for example, he or she is convinced that the Warren Commission erred in saying that Lee Harvey Oswald assassinated John F. Kennedy or that Richard Nixon did nothing criminal. My classes have tried both cases. But it takes the whole class, not just one student, for a trial to work. The student and I have to explain the argument to the class and convince them to do a trial. I try to remain low-key and let the student sell the idea. If the class votes to do a trial after an open discussion, great.

SETTING UP THE TRIAL

You need to set up teams of attorneys. They decide how the case will be presented and what witnesses will be called. I try to select capable students who are really interested in the case. Attorney teams can vary in size; I use two or three people to a side. Attorneys have to do a lot of research, prepare witnesses, and present the case in court. Some teams split up the work, one person per task.

Witnesses are important for a successful trial. Attorneys can provide them with information to use for their testimony or tell them where to look for it. In our Oswald trial, the most important witness was an investigator for the Warren Commission, played by the student who asked for the trial. He needed little coaching. In other cases, after attorneys convinced a classmate to be a witness, they gave him or her a volume of the Warren Commission testimony. It was easy for student witnesses to figure out their testimony when they had the real witness's testimony.

In the Nixon case, we had transcripts of all kinds of real-world testimony, not just a formal report. The biggest problem was selecting among so much material.

This preparation can take three or four days. Because the attorneys have to confer with each other and with the witness, usually some class time has to be available. Since only a fraction of the class, six attorneys and perhaps five witnesses, is involved in preparing the case, the question is what to do with the rest of the class. During the trial they will be the jury, but that doesn't solve the problem of dead time during preparation. My solution is to assign them a short research paper or an oral report on topics in the unit that do not pertain to the trial. This motivates some of them to volunteer as witnesses, because the choice is not loaf or become a witness but work one way or another.

The Trial

The judge's role is fairly simple. I usually take it myself. This puts me in a position of authority and in a position to advise the attorneys. Besides, the judge sits at my desk. Prosecution and defense attorneys move student desks to the front of the room, and we place a chair near the judge's bench for a witness box. The jury remain in their regular seats, as do witnesses. The trial is ready to begin. It follows the outline of trial procedure on page 144.

After the judge calls the trial to order, prosecution and defense make an opening statement. These are usually fairly brief, but if length looks like a problem, the judge can inform attorneys that they have a time limit. Then, the prosecution calls its first witness. Sometimes the witness is well prepared and responds confidently to questions. Sometimes not. When a witness pleads ignorance, the attorney has been known to reply, "The answer's on page 279." And the witness hurriedly flips to the proper page in the reference and resumes testimony.

It's unlikely that a trial can finish in one period, so it's necessary to call a recess. I keep an eye on the clock, and as we approach the end of the period, I announce a recess until the next class meeting. Then we quickly rearrange the room. A trial can take from two days to a week, depending on how much research the attorneys have done or how long-winded they are. In most cases, other members of the team tell their colleague to get on with it, so the judge rarely has to intervene.

When witnesses on both sides have been questioned and cross-examined, prosecution and defense attorneys present their arguments. When the two cases have been spelled out, there should be summaries and a final effort to sway the jury.

The judge gives the jury their instructions, and they retire to deliberate. Usually I tell them to find the defendant either guilty or innocent by a unanimous vote. If they are divided, it's a hung jury. Our jury room is a corner in a nearby hallway. When the jury returns, they announce their verdict, and the court adjourns.

Debriefing

Now is the time for some questions. How do the attorneys really feel about the case? Was there some fact you wish you had on the record that didn't get there? Did witnesses have some things they wanted to say but were not asked? What kinds of things influenced the jury?

That was how I used to do classroom trials, and that's how I would still do one in the right situation. In the current mock trial movement, the legal profession is trying to encourage mock trials as competitive events. I called a nearby law school to ask about the available materials, and in a few days a fat envelope full of interesting things arrived: very complete and detailed explanations of trial procedure, at least by my standards, including diagrams of how a courtroom is arranged and handouts containing the instructions a judge would give a jury in a criminal case and in a civil case. In some cases, the witness's information is only a short paragraph. In others, the whole trial is scripted. My favorite case involved a frog accused of theft. Other cases covered assault and battery, jury selection, and a case with conflicting testimony.

My students are always intrigued by legal technicalities, such as when the police can search their cars, and I suspect they would find these cases fascinating. If you are interested in trying either type of trial, you should write to the address below. They offered printed material and even the chance to have a lawyer participate as a resource person.

Address: National Law-related Education
Resource Center
American Bar Association
541 North Fairbanks Court
Chicago, IL 60611

Name ______________________________

Date ______________________________

GENERAL OUTLINE OF TRIAL PROCEDURE

1. Selection of jury
2. Opening statement by attorneys for prosecution
3. Opening statement by attorneys for defense
4. Prosecution's evidence
   (a) Direct examination of prosecution witnesses by prosecuting attorney
   (b) Cross-examination of prosecution witnesses by defendant's attorneys
   (c) Redirect examination by prosecuting attorneys
5. Defendant's evidence
   (a) Direct examination of defendant's witnesses by defendant's attorneys
   (b) Cross-examination of defendant's witnesses by prosecuting attorneys
   (c) Redirect examination by defendant's attorneys
6. Opening argument for prosecution
7. Argument for defendant
8. Closing argument for prosecution
9. Instruction to jury
10. Verdict of the jury

Chapter XIX

Reading and Writing

"I signed my daughter up for your class, Mr. Scott, because I knew she would have to write and work her tail off." That was the opening line of a half-hour conversation about our school. I listened a lot. There were several other interesting ideas, but my strongest impression was that it's nice to meet a parent who wants her child to write and to work hard.

In those days "back-to-basics" reading and writing was in, with admirable articles on the general importance of reading and writing, some elegant theories on how reading and writing relate to teaching social studies, and even a few articles on methods for teaching reading and writing in social studies classes. There was relatively little on how teachers could manage the extra work involved. "Back to basics" is no longer a hot topic, but for many parents and teachers, reading and writing remain very important. This chapter spells out how to give students some practice and instruction in reading and writing without working the teacher's tail off. The assignments are not always quick and easy, but they are reasonable and practical.

Quizzes as Practice Writing

For as long as I have been teaching, I have used pop quizzes. They serve a number of purposes: They encourage students to do the reading, to remember names of important people, and to understand important ideas. They stimulate class discussion and provide a writing exercise that requires almost no time to correct.

In textbooks a list of people to know and terms to understand usually follows the reading. It has names like Robert E. Lee and terms like *manifest destiny*. On the first day of school, I make it clear to my students that I reserve the right to give a quiz on any daily reading assignment, asking them to identify the people or define the terms. By the end of the first week of school, each of my classes has taken such a quiz. After that, I try to give a quiz frequently enough so it's a possibility students have to consider. This usually averages out to a quiz every two weeks. I do not give quizzes to all classes the same day. If my fourth hour has a quiz, the grapevine usually gets the word to at least some of my eighth-period students. So I will quiz my fourth period on Monday and my eighth period on Wednesday, or some similar pattern. This also means I don't have a huge set of papers to work on, just those from one class.

Giving the quiz is simplicity itself. I just say, "Please clear your desks of all textbooks and notebooks. Take out a sheet of paper." Then I write on the chalkboard, "Identify, explain, or define in the context of this reading . . .," and list five people or terms from the reading. I give students about five minutes to write their answers. I do not require that answers be in complete sentences. After the students have had enough time, I say, "Pass your paper to a sympathetic-looking neighbor." These neighbors do the preliminary corrections. I tell them, "Please sign your name at the bottom of the paper. Put 'corrected by' and your name." This helps students have a responsible attitude about correcting the papers.

Then the class discusses each item, and we arrive at right answers. Each correct answer is worth two points and the quiz is worth ten points. Sometimes, a partially right answer is worth one point. If there is a dispute about an answer, I have the correcting student note "re-read" at the top of the paper, and I check that answer when I record grades. After the papers have been scored and returned to their owners, I collect them for recording. I do not have to read them all because the students have done a good job in most cases. I do read a sample just to check on the accuracy of students' correcting. I always read the papers marked "reread."

I use a more elegant variation of the quiz as a writing exercise in a current events quiz. In that case I ask what the most significant news story of the past twenty-four hours has been, why it is important, and allow students five minutes to write a one-paragraph answer. I count down the minutes for them. This assignment requires that they write sentences, and the best answers will, in fact, be a paragraph.

Because of the time limit, the answers are short enough so I can read and grade each paper quickly. That means I can give quizzes frequently, thus forcing my students to write often. A variation would be to ask students to write a single sentence that summarizes the main idea of the day's reading. This would be a double-barreled exercise. It would encourage them to read for the main idea, and it would give them practice writing a clear sentence. I think I'll try that next year.

Student Papers

At the end of each semester, you see them—teachers staggering around with stacks (or boxes) of students' research papers. Those things can be monsters. My son once presented a biology teacher with a paper 110 pages long. How long would it take to read such lengthy work from a whole class? Many weekends and evenings.

There has to be a better way, and there is: the short paper, usually two pages. A teacher can read a set of two-page papers from a class in about an hour. I have done it hundreds of times. I find that my students do more research and writing for a series of two-page papers than for a big term paper. Consider the arithmetic. If you assign a 2,000-word term paper, that is about 10 pages. If you assign a two-page paper every three weeks instead, that works out to be six papers a semester using an 18-week semester. The total is 12 pages of writing in six different sessions. We know that students tend to write their papers in one last-minute rush. Which do you really think is more likely to improve writing skills? They probably also do the two-page paper in a last-minute rush, but they have six such experiences, and the two-page paper is less likely to keep them up until 2 A.M., and less likely to keep you or me up until 2 A.M. correcting.

Essay Tests

I have to confess to a powerful bias on this subject: I think essay tests are the best. I distrust multiple-choice and other objective types of questions. I have heard the lectures and read the books about how scientific objective testing is, but I still feel as if it's based on voodoo. With an essay test, I feel as if I can look inside my students' brains. It's not always a pretty sight, but I find out what's going on in there. Besides, essay tests are another way of having students practice writing.

There are times when giving essay tests may not be practical, but even then a little ingenuity will help. Let's assume that you are teaching five classes of 30 students each. This is a typical load. Unfortunately, it means 150 papers to correct. But why do all classes have to take the same test? Perhaps four classes could take the short-answer test and the fifth class do an exam that is half short-answer and half essay. Each time you give a test, a different class could have the essay. A rotating system like this cuts down the workload while giving students some essay-writing experiences. I used a system like this for a while.

A half-and-half system works well, too. For several years I gave tests made up of 30 multiple-choice questions and an essay. The mix of objective and essay seemed neatly scientific, and reading the essays was more manageable.

For the last several years, my unit tests have been all essay. Recently my students and I worked out a variation that we both like, the mid-unit test. I give a test about halfway through the unit, but it only lasts half a period. My students like it because they have to master only half as much material. I like it because I suspect that they study more and because I read half as much material at one time.

Let's look a the mechanics of giving essay tests. Most teachers believe that students should not know the questions in advance. This leads to a guessing game: Students try to anticipate questions; last year's students pass hints to this year's students, and the first period tells the fourth period. This is silly. If we have an idea or a set of facts we want students to master, why hide it? I hand my students the test questions at the start of each unit. From then on, they know what to study. I list perhaps four questions and select one, but they know the complete list of possibilities. Such certainty, I believe should encourage them to study.

The question themselves should be clear. This seems obvious, but I find writing clear essay questions quite difficult. I used to rely on "discuss" as the key to questions. Now I ask students to take a stand and provide examples to support that position. For example, I ask students to "Name the cause (or causes) of the Civil War and give a total of four examples showing the cause (or causes) in action." Students are not graded on the causes they name, but on the accuracy, completeness, and appropriateness of their examples. The other question I use regularly starts, "Trace the . . ." It is usually a straight recall question, focused either on lecture or a reading.

The simplest way to make questions clear is to explain what each calls for. Students often wonder if "four examples" in the Civil War essay means four examples for the essay or four examples for each cause. I assure them that they need only four examples in the essay, but that these must be fully explained, factually accurate, and linked clearly both to the cause they choose and to the start of the Civil War. Questions that require students to recall facts rather than analyze and synthesize material can be defined by telling students in class what sources provide answers: "Question four is based on this lecture. The wiser heads among you will take notes," or "Question two essentially asks you to summarize this chapter. Read it carefully."

For grading an outline of the correct answer, worked out in advance, is very helpful assigning a point value for certain items on the outline is also useful. For example, I often ask that students trace major military, political and diplomatic developments from 1775 to 1783—essentially, write an essay describing major events in the American Revolutionary War. They tend to get very involved with Minute Men and Redcoats, but often forget to mention the Declaration of Independence, after we spent two days in class on it! This costs them three out of a possible 10 points. The outline with assigned point values makes it easy to keep that in mind. When reading essays I am guided by precedent. When a paper has been read and graded, the next paper is, in part, rated against it. One paper gets a B; another follows an identical outline with the same level of detail, so it gets a B. I have read that it's a good idea to stack papers in order of their grades and read them a second time. It sounds like a good idea: I must try it someday when I have lots of time.

Rewriting

Usually when teachers read any student writing, we are reading first drafts. We all know first drafts are rarely good enough. In recent years I have offered students the chance to rewrite quizzes, essay tests, and papers. In my current events quizzes I offer a retake option called the Kristin quiz, named after the student who talked me into the idea. When I return the graded quiz, I say something like, "You now have a week to do a Kristin quiz. You can come in after school or in my free period. Just tell me you want to take a Kristin quiz, and I will give you a piece of paper and you can write. If your quiz grade is higher, I will replace the grade in my book. You cannot lower your grade."

A few students opt for Kristin quizzes. The second time around they have carefully prepared answers and can usually raise their grade. The first time around, they had often simply not prepared a current event for class. One quiz usually takes care of that.

For essay tests I make a similar offer: rewrite within a week, raise your grade. In some cases the original test allowed students to choose among different questions, so I can use the ones they didn't choose for the retakes. But, some questions are on the test regardless, like the causes of the Civil War question. In that case I coach students on how to improve their responses: "You need another example. You gave me only three. I asked for four"; or "You need to show more clearly how your second cause helped lead to the war."

On two-page papers I offer a similar option. When I return the papers, I put on the paper a little list headed "to get an A you need to . . ." and number each change I want. Rewrites are due one week after the paper is returned. Students may turn them in anytime before the deadline.

Please note that not all students take advantage of these rewrite options, but conscientious students at least have an opportunity to correct mistakes and be rewarded.

Reading

Let's start with a definition. Remedial reading is instruction that requires advanced training; so does speed reading. If I suspect a serious reading problem, I get word to the counselors or reading specialist. My business is to improve reading skills of a normal group of teenagers, principally by assigning readings and making sure students do them. If students are given a reading assignment and are not required to demonstrate that they have done it, they will not do the assignment, so I always try to couple reading assignments to a quiz, paper, or test. This gives me a certain reputation. My school librarian told me about one of my classes working in the library while I was gone for the day. They had a library reading assignment, and, as a student observer commented, "They really have to do it. That's Mr. Scott's class."

Reading Textbooks

Textbooks are the tools of the student's trade, and they should learn how to use them right. Most textbooks have built-in study aids, which I point out early in the year and explain how to use. I give students an incentive to use the study aids in their texts: "I use these study guides in making up your tests. When reading essays based on the reading, I look to see if you include all sections of the chapter and important material indicated by dark print. When I do multiple-choice tests, I write questions based on dark print, illustrations, and discussion questions. You know your test will include selected terms and people from the section survey." When I return tests, I point out again how students should have used the study aids. Some students take the hint and keep a systematic notebook of key ideas. My favorite example of a student who got the point is Diane. I was returning a multiple-choice unit test, and when I handed Diane her paper, I commented, "Not quite as good as usual." Diane looked a little uncomfortable and muttered something about, "Not bad, considering," I replied, "You don't have to tell me, but I'm curious about what's so unusual." Diane looked a little sheepish and then told me, "I didn't have time to read the assignment, so I just read the dark print." Her grade was B. Not bad, considering, indeed!

Book Reports

With so many "outlines" and "notes" available, not to mention summaries on book jackets, getting students actually to read the book is a challenge. I'll suggest a few ideas that work well for me— not perfectly, for no system is perfect.

Paperback books are a great invention. They make it possible for schools to own several copies of a supplemental book at a reasonable cost, and most schools are comfortable with buying paperbacks. Perhaps you can select a good paperback and get a set of 30 to keep in the classroom. It's wise to read the book before ordering it, to check it for student appeal and reading level. Take a few notes on content. If the school is not able or willing to purchase paperbacks, perhaps students can be asked to purchase books. The big advantage of having all students read the same paperback is that the teacher has to read only one book to evaluate student work. The big disadvantage of using a single paperback title is that students can share their reviews. Teachers should be alert for this. It's a wise precaution to reserve the right to ask each student a few questions about the book, using your notes on content for reference. Perhaps the simplest way to verify that a class has read the book is a short quiz. Use alternate forms if you have more than one class.

I never accept a book report on a book I haven't read. After twenty years of teaching, I've read a lot of books, so that is not too severe a restriction. Also, I sometimes make an exception and take a book home over the weekend to preview it if a student asks for permission to use a book that I haven't read that looks acceptable.

My instructions on book reports always remind students that their reports have two purposes: to show me that they have read the book and thought about it. I also remind them that with a good book review is coherent, a thesis statement supported by related information. It sounds a little as if I'm teaching writing. Could it be that teaching reading and writing are related?

A Reading Break

Did you ever notice the books your students carry? I'm not referring to textbooks. I have noticed that really good students carry pretty trashy stuff, pure entertainment, *World's Worst Jokes*, or the like; average students carry serious stuff, assigned by English teachers; and the poor students don't carry books. This ought to tell us something. Poor students never read unless they have to, and then reluctantly. Average students read just what they have to, but no more. Good students not only read what they have to but read for fun, too. These observations mean that as soon as most of our students get out of school, they will stop reading.

In the 1980's several schools in my area and, for all I know, across the country, were using a program to make students read for pleasure. It's relatively simple and cheap to implement. My observations are those of a teacher in the program. Experts on the program have written on it and lectured on it. See the bibliography for details. The idea is simply to take a 15-minute period in the school day and have everybody read. In some schools that means every student and every adult in school: janitors, secretaries, teachers, and the principal. In my school the teachers read, but the support staff does not, and one disciplinarian remains on duty. There is an eerie silence throughout the school.

There are all kinds of practical questions about initiating such a program. How to make time for it? Where to obtain reading materials? Could everybody be required to read? Would the staff cooperate? What would students think? What changes would it make in attitudes toward reading and reading abilities?

Reading was scheduled in the middle of the morning, just before the third-period classes. It lasted for fifteen minutes. We found fifteen minutes by shaving two minutes off each of our eight class periods. If reading is important, you can spare two minutes per class for it, right?

Materials require a little more ingenuity. Funds were limited. We spent some money on magazine subscriptions to appeal to people with little interest in academic subjects. We asked for book donations, and predictably we received lots of paperbacks—no joke books or romances, but lots of copies of *The Little Prince* and *Patterns of Culture*.

The theory was that students would bring their own reading material for the reading session. Any experienced teacher knows how often that happened, less than half the time. Most teachers put together a collection of books and magazines in their rooms. The most sought-after item in my room was the newspaper. In addition, I had a selection of magazines from our local public library and some paperbacks from the school stockpile. The magazines came from the magazine exchange: People contribute magazines to boxes located in our library entry. They're free for the taking, with no checkout and no obligation to return. It's a great system. In spite of my efforts to keep the shelf stocked with a variety of material, students wanted more.

Getting everybody to read is a very tricky business. In some classrooms at first, students studied or just chatted. Teachers put pressure on their peers, and behavior was eventually corrected. In my room the problem was to make sure that students were not reading material assigned for their classes. For the first day or two, I gave up my own reading time to patrol. I told people reading textbooks to put them away and read a magazine. I pushed people to bring material they liked. Strict silence was the rule. Any student violating it received a reproving look, followed by a whispered reminder. My students were in an academic elective, so it didn't take much effort to get them settled down to read.

In general, the staff was enthusiastic about the program and worked to make it succeed. They were polled on it every year with a favorable response. The students were also polled. Their responses were quite favorable as well, the most common complaint being that the reading period was too short.

The most difficult thing to pin down is changes in attitudes and abilities. We did not set up a study to measure scientifically, but anecdotal evidence was encouraging. My favorite example is John. He matured late and had a persistent problem with reversing "b" and "d." His elementary school reading development was slow. When he enrolled in junior high, the school had a reading program like the one just described. His parents encouraged him to read paperback adventures. He became a Louis L'Amour fan, devouring L'Amour's books as fast as they came out. He is now an excellent reader; he knows how it happened. "You know, Dad," he told me, "that reading program was just what I needed."

BIBLIOGRAPHY

Beyer, Barry K. *Back-to-basics in Social Studies.* Social Science Education Consortium, Inc., 1977. (Includes a discussion of "What Are the Basic Basics?" and a set of guidelines and techniques with some sample lessons teaching reading, writing, and thinking with content.)

Daddow, Kirk, and John Forsman. *"Writing in Every Subject: A Shared Responsibility."* Unpublished, 1980. (This paper, prepared and presented by two staff members at Ames High School, provides several ideas about the place of writing in American history.)

Finkelstern, Joseph. "The Two-page Essay." *Social Education,* 26 (October, 1962) 301-302. (This is the article that started me thinking about using the two-page paper.)

Quinn, Sandra L., and Sanford B. Kanter. *How to Pass an Essay Examination.* Dubuque, IA: Kendall/Hunt Publishing Co., 1982. (A college-level book full of very practical advice on writing essay exams: clear, direct, and honest. I experienced a shock of recognition reading about student con games. Teachers should read this, and they may want to share parts of it with their students.)

Social Education, 43 (March, 1979). (This issue contains a series of articles on writing in the social studies.)

Wolfe, Denny, and Robert Reising. *Writing for Learning in the Content Areas.* Portland, ME: J. Weston Walch, Publisher, 1983. (This teacher's handbook contains valuable suggestions for using writing in five subject areas, including social studies.)

CHAPTER XX

WHEN THE TEACHER OR THE STUDENT IS AWAY FROM SCHOOL

I know I'm in trouble as soon as I realize it's morning. My body is one large, dull ache with a few little sharp ones thrown in for variety. There is a small bonfire in my throat. My wife, ever cheerful, asks, "How are you this morning, dear?" I suppress the great wave of hostility that I feel toward all cheerful, healthy people and merely groan softly. It hurts to groan, even softly. I'm in real trouble. I lurch to the phone and dial the number of my school. The cheerful voice of our associate principal comes on. Boy, do I hate cheerful people today! In as few words as possible, I tell him that I have the flu and will need a substitute. He responds that the substitute will call me in a few minutes to discuss my lesson plans. I respond something like, "Sure . . . lesson plans," and we hang up.

You have probably experienced a similar morning if you have been teaching for many years. Most of us miss a day or two due to illness. There are a few useful and educational strategies for a class taught by a substitute or by a teacher who is too hoarse to talk. Let's look at a few possibilities.

There are a few obvious ideas that I consider cop-outs. Having students read their textbooks or go to the library are often in this category, unless there is a special reason for students to read the text or work in the library during class time. Otherwise, I suspect that time is wasted.

When I have to be away for a meeting, I often set up games to be played the day I am gone. If you suspect an illness is coming on, you can do that, too. If I'm lucky, some of my classes will already have a game in progress. If they are playing the Stock Market Game, for example, it can be adjusted to the situation. Ordinarily, the game runs for just 10 or 15 minutes at the start of class. In an emergency, the substitute can accelerate the schedule and run three or so "days" of the game in one class period. This game demands little of the person running it, and it's entirely manageable for a substitute, especially if he or she has read the chapter on the game. In fact, if a copy of this book is kept handy for the substitute, there are a number of single-period games that a substitute could manage: Mini-States, the Congress Game, the Cattle-Town Game, the Railroad Game, and 1898. Some of these require advance planning, but once you have the handouts in your file, they can be done quite easily.

The ideal class activity for when the teacher is gone or unable to talk is a test. Since I give very frequent tests, there are often one or more of my classes taking a test when I'm gone. This is partly because tests can be scheduled, so I will move a test either ahead or behind scheduled time to cover a day I know I will not be in class. Since I reserve the right to give a quiz on any day, a quiz can be the opening event of the lesson plan. (See Chapter XIX, page 137, for a description of how my quizzes work.)

Next we get into the obvious area of videotapes. If available, they're lifesavers, but they're also a trap. Students assume that a video is a good chance to relax, nap, or daydream. The obvious way to beat this is to ask each student to write a short paragraph summarizing the video. The equally obvious disadvantage is the pile of papers awaiting you. To reduce that pile to a fraction, ask the substitute to put students in groups of four to six and have each group collaborate on a paragraph. All students in the group must sign the paragraph, and you need to grade it. You don't have to make it an important grade, but students have to know you do check it.

With a video or other audiovisual aid that you have already used, you can have a really elegant lesson plan ready for your substitute. When you preview the video, take notes on ideas, information, and discussion questions. After that, prepare a transparency master and include the title of the video. (I even include the code number our supplier uses to be sure that I have the right video.) Then, put down the discussion questions, the questions about important facts, or whatever else you want your students to do after they watch the video. Make your transparency and project the questions at the start of class. While you are setting up the VCR, students can note down the questions. After running the video, you can project questions again and lead students through them. After class, carefully file the transparency. You don't need an overhead projector to make the questions work. Notes on the chalkboard are just as good. Either way, the plan will be available next year, even if a substitute is sitting in. A few of these filed away can take care of a lot of emergencies.

Students can also tape oral reports scheduled while you are gone, and you can grade them from the tapes. If they are not using videotapes, be sure that they identify themselves clearly at the start of their reports. I've spent some very frustrating hours listening to unidentified students.

Now let's consider the worst case possible. No tests scheduled, no movies, nothing; you were just going to discuss the text. That's a desperate situation, but not hopeless. I use this technique two or three times each year, when my throat is too sore to talk much or when a substitute takes the class. The assumption is that the class has been assigned a typical text reading and students have their books with them. I divide the class into three work groups and assign each group a portion of the text assigned and a space at the chalkboard. The task is to prepare an outline of the reading and have one member explain it to the class. If you have an overhead projector, you can issue an acetate sheet and a grease pencil to each group so they can project the outline instead of using the chalkboard. Where chalkboard space is very limited, the group can prepare an outline on paper and one member can write it on the board while another member talks about it. The next group can erase that outline and put theirs up. If you are gone, have each group turn in their outline, on paper or acetate, with all the members' names, so you can grade it. Setting up groups and explaining the task takes two to five minutes. The outlining itself takes 10 to 20 minutes. The teacher or substitute needs to monitor progress and keep students focused on the job. Reporting takes 10 to 20 minutes, depending on the amount of comment the teacher wants to interject, students' questions, and class time available. The experience is not quite so valuable, perhaps, as a discussion led by a skilled teacher, but it's not bad.

Extended Student Absences

Today's student is highly mobile. It seems to me that nearly every year one of my students zips off to a foreign country, like Nigeria or even Texas. Sometimes it's a temporary relocation, in which case the family may want the student to return to graduate. I have twice arranged programs for students who were living temporarily in foreign countries. The students made substantial progress during that time.

But first let's consider the case of a student who was not quite mobile enough. Jorge was hit by a car and sustained a broken thigh and a full cast. He was unable to attend school for a few months, so the school came up with the following arrangement.

Jorge was immobilized and confined to bed, but he was in town. Our school administrators arranged for a phone hook-up. It was a little gray box with a cord to plug into a jack in the classroom, and it worked like a speaker phone, that is, it picked up anything said in the room during class discussion so Jorge could hear it. Jorge had a button on his device. When he wanted to contribute to the discussion, he pushed his button, and we heard a click on our end. I called on him much as I did any other student. He had textbooks and assignment sheets at home and managed to keep up with the class quite well. To move the little gray box from class to class, we simply had a reliable student carry it from one room to the next. We called it "carrying Jorge around."

This was a few years ago. Recently, I talked with a representative of a major phone company and she told me there is now a very sophisticated collection of equipment for videoconferencing. If both the student and class have a robust desktop computer modified with a video card, a device to send pictures digitally, and a video camera, they can hook up on an ISDN line and be able to see and hear both ways. The picture is not quite television quality, but the picture changes 15 times in a second. The student can call up the classroom and see whatever the camera is aimed at. With videoconferencing, the student could hear and see lectures and even take part in small-group work. Students can read handouts and type answers from home into the classroom. If the teacher is showing a video, the VCR can be connected to the computer, and the student can watch the video on the computer screen at home, with full sound and pictures, just as the rest of the class is seeing and hearing it.

This can be costly. The equipment described above is about $1,800 and the ISDN line is $60 a month, but in some situations that looks reasonable. Consider a rural school with three students who want to learn Japanese. This system would allow them to enroll in a university hundreds of miles away.

In some cases videoconferencing is better than actually having the student in the classroom. Students with disfiguring injuries or hair loss from cancer therapy who are embarrassed about personal appearance can use "ap guards," cartoon images that appear on the classroom screen to represent the sick student.

A school with several students who cannot attend school can do bridging. Bridging is much like a conference phone call. Up to 50 students can dial in simultaneously in a class. The same system would let a class take a call from a distant guest speaker. A class studying the American Revolution could talk with a museum director, and all the students in the class and at home could participate. The possibilities are limited only by your imagination.

Jorge's situation was an extreme one. Usually absences last only for a few days. If a particular class event is very important, it's a good idea to tape it for those who miss it. I once taught a modern problems course without a text and taped all my large-group lectures.

Enough of this hometown stuff; let's get to those students in foreign lands. This takes quite a bit of organization and careful planning, but if the student really wants it, the system can work. In each case I have known the student quite well. There also has to be a responsible adult on the student's end who can manage the mechanics.

The process usually begins with a conference involving the teacher, the student, and the adult who will supervise the student. In that conference we work out the details of the course. The student takes a package overseas, including textbook, assignment sheets, and project materials. The adult has a set of envelopes containing tests. When the student is ready for a test, he or she can ask to take the test. The adult then gives the student the envelope and supervises the test, which is mailed to me for grading. (If you want more detail on how I do tests, see Chapter XIX.)

Arranging a program for a student going abroad is not easy. Because we cannot assume access to a library with American history materials, the package must be self-contained. I use a series of book reports for projects, based on paperbacks that will be helpful for understanding an important part of the unit. These books have to be interesting and on the appropriate reading level. I usually ask for five-page book reports.

Once we have worked out all details and everybody concerned understands them, we write a contract, including standards for grades. It is signed by the teacher, the student, the student's parents, and an administrator. This contract is made partly to furnish everyone with a record of what is expected and partly to impress the student with the idea that he or she is now committed. (See sample contract on page 147.)

Once the trip begins, all I do is wait. When an envelope comes in containing a book report or a test, I grade it, write extensive comments, and send them back to the student. In one case I tape-recorded my reaction to the work and gave a little pep talk about the next unit. I thought that the sound of a familiar teacher's voice might serve as a motivating factor.

Living abroad is a powerful distraction, and correspondence study takes a lot of self-discipline. None of my overseas students completed the course while they were abroad. They made progress but had to finish when they got home, but that only took a few weeks.

More recent technology might make it easier. I suspect that today we could combine e-mail, the World Wide Web, and videoconferencing to keep students abroad motivated. Using videoconferencing overseas is probably prohibitively expensive, but a few face-to-face tutorials could really motivate students to fulfill that contract.

I have also had a number of students who did similar contract study while enrolled in and attending our high school. They were usually bright, mature people who could find no other way to work the class into their schedule. We usually scheduled meetings once a week, and students were able to complete the course satisfactorily.

From all of these experiences, I conclude that for most students the best learning takes place in a classroom with a healthy teacher. A few exceptional students thrive on other systems, and with little creativity, teachers can provide reasonably acceptable classes while they are absent. For times when either the teacher or the student can't make it to class, it's best to be prepared with some ideas and strategies.

SAMPLE CONTRACT

UNIT FORMAT—INDEPENDENT STUDY
AMERICAN HISTORY

I. Unit reading assignments
- Todd-Curti, *Rise of the American Nation* (pages listed).
- A book related to the unit. I will either name it or give you a list of possibilities and let you have your choice.

II. Unit Evaluation
- A one-period essay test graded against an outline of the text reading including
  - (a) A list of ten people, events, or concepts taken from the section survey in the text.
  - (b) A question asking you to summarize one chapter in the text reading in the unit.
  - (c) A question based on a problem of interpretation or judgment in the unit. This may also include your book reading.
- On the book, you may be asked to do one of the following, depending on the unit.
  - (a) Write an essay exam on the book stressing the main ideas of the book.
  - (b) Write an out-of-class essay or on the book. Often this will involve a question I give you.
  - (c) Use the book in answering essay question under A above.
  - (d) Take an informal oral examination over the book.
- All grades will be letter grades.

III. Semester examination will be a long multiple-choice examination based on the text. It will be 1/5 of the semester grade.

IV. Grading standards for the course grade
- Honors "A" must deliver at least 8 out of 9 units on an A level.
- "A" if a majority of the unit grades are A level and if numerical average of the unit grades works out to above 3.5.
- "B" numerical average 3; "C" numerical average 2; "D" numerical average 1.
- Failure if all units are not completed or numerical unit grade is under .5.

Student ______________________

Teacher ______________________

Parent ______________________

Administrator ______________________

Share Your Bright Ideas with Us!

We want to hear from you! Your valuable comments and suggestions will help us meet your current and future classroom needs.

Your name______________________________ Date______________

School name______________________________ Phone______________

School address______________________________

Grade level taught________ Subject area(s) taught______________________ Average class size______

Where did you purchase this publication?______________________________

Was your salesperson knowledgeable about this product? Yes_____ No_____

What monies were used to purchase this product?

___School supplemental budget ___Federal/state funding ___Personal

Please "grade" this Walch publication according to the following criteria:

Quality of service you received when purchasing	A	B	C	D	F
Ease of use	A	B	C	D	F
Quality of content	A	B	C	D	F
Page layout	A	B	C	D	F
Organization of material	A	B	C	D	F
Suitability for grade level	A	B	C	D	F
Instructional value	A	B	C	D	F

COMMENTS:______________________________

____

What specific supplemental materials would help you meet your current—or future—instructional needs?

____

Have you used other Walch publications? If so, which ones?______________________________

May we use your comments in upcoming communications? ___Yes ___No

Please **FAX** this completed form to **207-772-3105**, or mail it to:

Product Development, J.Weston Walch, Publisher, P.O. Box 658, Portland, ME 04104-0658

We will send you a **FREE GIFT** as our way of thanking you for your feedback. **THANK YOU!**